CIRCUIT ANALYSIS

CIRCUIT ANALYSIS

Leonard J Tung
Bing W Kwan

Florida State University
FAMU-FSU College of Engineering
Department of Electrical & Computer Engineering

World Scientific
New Jersey • London • Singapore • Hong Kong

Published by

World Scientific Publishing Co. Pte. Ltd.

P O Box 128, Farrer Road, Singapore 912805

USA office: Suite 1B, 1060 Main Street, River Edge, NJ 07661

UK office: 57 Shelton Street, Covent Garden, London WC2H 9HE

British Library Cataloguing-in-Publication Data
A catalogue record for this book is available from the British Library.

CIRCUIT ANALYSIS

ISBN 981-02-4051-1

Printed in Singapore by Mainland Press

CONTENTS

PREFACE

This volume is intended as a textbook for a first course in Electrical Engineering. The materials included are divided into ten chapters that can be divided into two parts with five chapters each, for a two-semester coverage.

The first part with the first five chapters deals with circuit elements, resistive circuits, circuit theorems, circuit topology, and the state variable method. The presentation of the state-variable method is a special feature. The authors believe that the natural way to analyze RLC circuits is to use the state-variable method rather than second- or high-order ordinary differential equations. By choosing capacitor voltages and inductor currents in an RLC circuit as state variables, the so-called state equations can be systematically obtained through network topology. Of particular interest is the approach employing Thevenin's Theorem and Norton's Theorem to find the state equations without using circuit topology. Furthermore, the state equations can be constructed by using the DC analysis of a SPICE program.

The last five chapters make up the second part of the book. This part covers sinusoidal steady-state analysis, three-phase circuits, two-port networks, Fourier Analysis, and Laplace Transform. Great effort has been devoted to presenting the subjects of Fourier Analysis and Laplace Transform with examples of practical circuits. Thus, we hope that the reader will be better motivated to learn rather abstract concepts such as complex frequency and frequency response.

The authors are indebted to Dr. Rodney Roberts, a true colleague, for his suggestions and comments that have made this book more comprehensive and self-contained. The authors would also like to thank Ms. Yubing Zhai, Mr. Steven Patt, and World Publishing Co. for their patience and persistence that have made possible the publication of this book.

Chapter 1
Circuit Elements

What is a circuit? Generally speaking, a circuit is a collection of circuit elements interconnected through wires in a specific manner. Circuit elements are basic circuit models for electrical devices with two or more terminals, made of wires, for connection with each other. In order to analyze a circuit, we need to understand some basic electrical phenomena and establish certain terminology first.

1.1 Circuit Quantities

There are five basic electrical quantities of interests in a circuit. They are: charge, current, voltage, energy, and power. Among these five quantities, current, voltage and power are more important in circuit analysis.

1.1.1 Charge and Current

In the physical world, there are two kinds of charges: positive and negative charges. Charges are measured in coulombs (abbreviated as *coul*) and they are quantized. The tiniest amount of charge is that of an electron which is about 1.60×10^{-19} *coul*. Since electrons are negative charges, an electron is denoted as

$$1 \quad electron = -1.60 \times 10^{-19} \quad coul$$

When a group of charges is forced to move about, it gives rise to an electrical current. In a circuit, an electrical current is normally confined either in a piece of wire or in a device. In a particular piece of wire, the current through the wire is defined as follows:

Given a cross-sectional area of a wire, the amount of charges in *couls* that flow from one side of the cross-sectional area to the other per second defines the measurement of the current in *Amperes* (abbreviated as *Amp* or *A*) of the chosen direction.

The current so defined can either be positive or negative in value. When the moving charges are negative in nature, the current flowing in the direction of these moving charges is negative in value according to the definition. A current is also directional. Usually, the direction of a current can be assigned arbitrarily. Its value once found, whether positive or negative, would indicate the direction of the actual current. A current flowing in a particular direction and with a negative value is equivalent to a current flowing in the opposite direction and with a positive value of the same magnitude. Let's examine the circuit in the following figure.

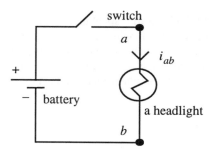

Figure 1.1 The circuit for a headlight in a car

The circuit shown in **Figure 1.1** is found in a typical car. A headlight in a car can be turned on through a switch. Once the headlight is turned on, the car battery forces a current through the headlight that in turn transforms electrical energy into light and heat. Omitted from the circuit is a fuse that protects the circuit from overload. The current flows from the positive terminal of the battery through the light to the negative terminal of the battery. Inside the battery, the current flows from the negative terminal to the positive terminal. What actually happens in a circuit like this one is complicated. However, we are going to concentrate only on the electrical phenomenon. The actual chemical process in the battery and the actual energy transformation in the light are beyond the scope of this book.

Let q denote the net charge that reaches terminal a of the light from the positive terminal of the battery. The rate change of q gives rise to the current. Therefore, we have

$$i_{ab} = \frac{dq}{dt}$$

1.1.2 Energy and Voltage

It takes energy to move charges around. In the circuit of **Figure 1.1**, the energy is originated from the battery through chemical reactions. It takes about twelve *joules* of energy to move one *coul* of positive charges inside the battery from the negative terminal to the positive terminal. Accordingly, the voltage of the positive terminal of the battery with respect to the negative terminal is said to be twelve *volts* (abbreviated as *V*). A voltage of one *volt* indicates that one *joule* of energy is needed for moving one *coul* of positive charges. In any circuit, the voltage of a particular point can be defined with respect to any other point in the circuit as follows:

The voltage of point *a* with respect to point *b* in a circuit, denoted as v_{ab}, represents the amount of energy in *joules* needed to move one *coul* of positive charge from point *b* to point *a*.

At this moment, it is important to point out that the voltage so-defined is independent of the path taken by the charge from point *b* to point *a* in a circuit. For the battery, we have

$$v_{+-} = 12 \quad V$$

Once a positive charge is moved to the positive terminal of the battery, it possesses electrical energy to move through the rest of the circuit. When this positive charge moves through the headlight and back to the negative terminal of the battery, it gives out all the electrical energy acquired from the battery. Consequently, we write

$$v_{-+} = -12 \quad V$$

The negative value indicates that the charge gives out instead of acquiring electrical energy.

In **Figure 1.1**, the "−" terminal of the battery is connected to the terminal *b* of the headlight through wire which is made of a conductor. The voltage between these two points is approximated by zero. Similarly, the voltage between the "+" terminal of the battery and terminal *a* of the headlight is also approximated by zero.

1.1.3 Power and Passive Sign Convention

In **Figure 1.1**, a current flows continuously from the positive terminal of the properly charged battery through the light to the negative terminal when the switch is turned on. Let *i* denote this current measured in *A*. For every second, there are *q* *couls* of charges that reach the positive terminal of the battery. For one *coul* of charges started from the positive terminal, the charges lose twelve *joules* of electrical energy to the light when the charges eventually gets to the negative terminal of the battery. Hence, the light gains twelve times *q* of electrical energy in joules every second and turns this energy into light and heat. We say that the light consumes (or absorbs) twelve times *q* of electrical energy every second. The rate at which the electrical energy is consumed by the light is known as the power consumed by the light. Let *p* denotes this power. We have

$$p = \frac{d(12q)}{dt} = 12\frac{dq}{dt} = 12i_{ab} = v_{+-}i_{ab}$$

where the subscript a denotes the point of the light connected to the positive terminal of the battery while subscript b denotes the point connected to the negative terminal of the battery. It is quite cumbersome to denote voltages or currents with subscripts. Using the polarity signs and the direction sign, we can easily identify a voltage or a current, respectively. Extracting out of **Figure 1.1**, the light is further identified with a voltage variable and a current variable as shown in **Figure 1.2**.

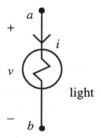

Figure 1.2 A circuit element representing the light in Figure 1.1

For the power consumed by the light, we have

$$p_l = v \cdot i$$

The power is measured in watts (abbreviated as W) if the voltage is measured in *volts* and the current is measured in *amperes*. In general, the power is a function of time t.

Using the voltage polarity and the current direction assigned (from "+" terminal to the "−" terminal through the light), we can represent the power absorbed by any element with the same product of the voltage and the current. This way of assigning the voltage polarity and the current direction for a circuit element is known as the passive sign convention. Unfortunately, it is not always possible to follow the passive sign convention as demonstrated in the following example.

Example 1.1. Assume that the current through the light in **Figure 1.1** is measured to be 0.5 A. Calculate the power absorbed by the light. What is the power absorbed by the battery?

Solution:

For the light, the passive sign convention is observed. Hence, we have

$$p_l = v \cdot i = 12 \times 0.5 = 6\,W$$

In the case of the battery, the passive sign convention is not observed. However, a current of 0.5 A flowing through the battery from the "−" terminal to the "+" terminal is equivalent to a current of − 0.5 A flowing through the battery from the "+" to the "−" terminal as shown in **Figure 1.3**.

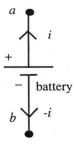

Figure 1.3 The power caculation of a battery

The power absorbed by the battery is therefore given as follows:

$$p_s = v \cdot (-i) = 12 \times (-0.5) = -6\,W$$

Saying that an element is absorbing − 6W of power is equivalent to saying that the element is supplying 6W of power.

1.2 Element Classifications

To simplify the drawing and the analysis of a circuit, electrical devices are represented by simple models called circuit elements. Circuit elements posses various characteristics. According to the characteristics, elements are classified according to the following categories:

1) Passive or Active Elements
2) Linear or Nonlinear Elements
3) Resistive or Energy-storage Elements
4) Voltage or Current Sources

5) Dependent or Independent Sources

These categories are discussed in the following sections.

1.2.1 Passive versus Active Element

From the electrical power $p(t)$ absorbed by an element, we can calculate the total electrical energy absorbed by an element at any instant according to the following equation:

$$w(t) = \int_{-\infty}^{t} p(t)dt$$

If the energy $w(t)$ is always greater than or equal to zero, the element is said to be passive. If not, the element is said to be active. In **Figure 1.1**, the light is a passive element while the battery is an active one. In addition, resistors, capacitors, inductors, and diodes are passive elements. Batteries and electric outlets are classified as active elements.

1.2.2 Linear versus Nonlinear Element

For a passive element, either the voltage across it is a real valued function of the current flowing through it or the current flowing through it is a real valued function of the voltage across it. If such a function is linear, the element is said to be linear. If not, the element is said to be nonlinear. A linear function $f(x)$ is characterizes by the following properties:

$$\begin{cases} f(x_1 + x_2) = f(x_1) + f(x_2) \\ \quad f(\alpha \cdot x) = \alpha \cdot f(x) \end{cases}$$

for any real constant α and any arguments x_1, x_2, and x. Resistors, capacitors, and inductors are usually modeled as linear elements. Diodes are nonlinear. In this book, we are mainly concerned with linear elements.

1.2.3 Resistive versus Energy-Storage Element

For a passive element such as a resistor, it is found that the power absorbed is always greater than or equal to zero at any instant. Such an element is said to be resistive. Diodes are also considered as resistive although they are nonlinear.

For a capacitor or an inductor, the power absorbed by the element at a particular instant can be negative. This indicates that the element is actually gives out electrical energy at that instant. The energy that the element gives out has been stored in the element ahead of time. Hence, this type of elements is known as energy-storage elements.

1.2.4 Voltage Sources versus Current Sources

Active elements are classified into two categories: voltage sources and current sources. A voltage source is an active element that supplies electrical energy at a specific voltage across two terminals. The terminal voltage of a voltage source can be constant like that of a battery. It can also be a specific time function like that of an electrical outlet. This type of sources is normally identified with a circle or a diamond shape with a polarity sign inside it as shown in **Figure 1.4**. The difference between these two shapes is to be explained in the next section.

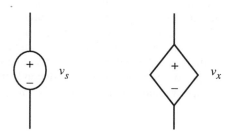

Figure 1.4 Symbols for voltage sources

The voltage v_s (or v_x) refers to the voltage of the positive terminal with respect to the negative terminal. In the case of a battery, the circle together with the polarity sign is replaced by two parallel bars of different lengths as shown in **Figure 1.1**. The longer bar indicates the positive terminal of the battery while the shorter bar the negative terminals.

A current source is an active element that supplies electrical energy at a specific current from one terminal to the other. The current through the element can be a constant or a time function. This type of sources is normally identified with a circle or a diamond shape with a direction sign inside it as shown in **Figure 1.5**.

The current i_s (or i_x) represents the current through the current source in the direction shown. A photo diode controlled by a light beam in the assembly of an automatic door can be modeled approximately by a current source.

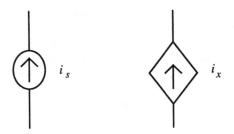

Figure 1.5. Symbols for current sources

1.2.5 Dependent versus Independent Sources

In **Figure 1.4** and **Figure 1.5,** both circles and diamond shapes are utilized to represent sources. Circles are normally utilized to represent the so-called independent sources while diamond shapes are reserved for the type known as dependent sources. A voltage source is classified as an independent one if the source voltage v_s is independent of any other circuit quantity. A current source is classified as an independent one if the source current i_s is independent of any other circuit quantities. Batteries and electrical outlets are modeled as independent sources. A source is classified as a dependent one if the source function (voltage for voltage source and current for current source) is dependent on another circuit quantity. Transistors and Operational Amplifiers are often modeled with dependent sources.

1.3 Resistors and Ohm's Law

Many devices that consume electrical energy can be modeled by resistors. For example, electrical lights and electrical heaters are normally represented by resistors. The characteristics of a resistor is best described by the so-called Ohm's Law.

1.3.1 Ohm's Law

Georg Simon Ohm studied the electrical properties of different kind of materials with various uniform cross-sectional areas and various lengths as the one shown in **Figure 1.6.** His findings can be summarizes as follows:

(a) The voltage v across a specimen is linearly proportional to the current i through it. This is known as the Ohm's Law

(b) The proportional constant, known as the resistance and denoted as R, is linearly proportional to the length l of the specimen and inversely proportional to the cross-sectional area B of the specimen.

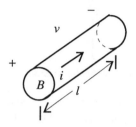

Figure 1.6 A resistor in cylindrical shape

Thus, we have

$$v = R \cdot i = \left(\rho \frac{l}{B} \right) \cdot i$$

The proportional constant ρ is known as the resistivity and is dependent on the type of material that the specimen is made of. If it attains 1 *volt* of terminal voltage when 1 *Amp* of current is flowing through the specimen, the resistance of the specimen is quantified as 1 *ohm* or 1 Ω. The equation above can also be written as

$$i = \frac{1}{R} v$$

The inverse of resistance is known as the conductance. Conductors usually have low resistance. Other types of materials, such as graphite, are utilized to attain higher resistance. Devices with the characteristics governed by Ohm's Law are called resistors.

1.3.2 Symbol and Power Consumption

The symbol for a resistor is shown in **Figure 1.7**.

The power absorbed by a resistor can be found as follows:

$$p = v \cdot i = (R \cdot i)i = R \cdot i^2 = v(\frac{v}{R}) = \frac{v^2}{R}$$

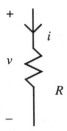

Figure 1.7. Symbol for a resistor

As shown in the equation, the power absorbed by a resistor is always greater than or equal to zero. In our daily life, incandescent lights and electrical water heaters can be properly modeled with resistors. As a matter of fact, any device that consumes electrical energy requires a certain kind of resistor (linear or nonlinear) as parts of the device model.

1.3.3 Special Cases

There are two special cases that are of importance in the circuit analysis: (1) the resistance is zero or (2) the resistance is infinity.

Case 1. Short-circuit:

When the resistance of a resistor approaches to zero, we say that the resistor is shorted as shown in **Figure 1.8**. The voltage across the two terminals becomes zero while the current through the resistor can take on any value only restricted by how the resistor is connected to other elements in a circuit.

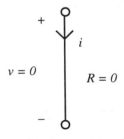

Figure 1.8. A short circuit

Case 2. Open-circuit:

When the resistance of a resistor approaches to ∞, we say that the resistor is opened as shown in **Figure 1.9**. The current through the resistor becomes zero while the voltage across the two terminals can take on any value only restricted by how the resistor is connected to other elements in a circuit.

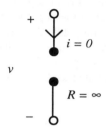

Figure 1.9. An open circuit

1.4 Energy-Storage Elements

There are two kinds of basic circuit elements that store electrical energy: capacitors and inductors. A capacitor stores energy through an electrical field while an inductor stores energy through a magnetic field.

1.4.1 Capacitors

The simplest capacitor consists of two parallel plates separated by a short distance d. The side view of such a capacitor is shown in **Figure 1.10**.

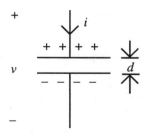

Figure 1.10. A set of two parallel plates

If the parallel plates are connected to a battery through a resistor, positive charges will accumulate on the plate that is connected to the positive terminal of the battery. This plate is denoted as the positive plate. Meanwhile, an equal number of negative charges will accumulate on the other plate that is connected to the negative terminal of the battery and is denoted as the negative plate. Right after the connection is made, it takes little energy to move charges of different kinds on opposite plates. As charges start to accumulate on both plates, it becomes more and more difficult to increase the amount of charges on either plate. Hence, the voltage v of the positive plate with respect to the negative plate becomes larger as time goes on. Eventually, current stop flowing while the amount of charges reaches a constant level on each plate. Let q denote the amount of charges accumulated on the positive plate. The charge q is found to be proportional to the voltage v across the plates. The proportional constant is denoted as C and is called the capacitance. Thus, we have

$$q = C \cdot v$$

The capacitance is found to be linearly proportional to the area B of either plate and be inversely proportional to the distance d between the parallel plates. Hence,

$$C = \varepsilon \frac{B}{d}$$

where the proportional constant ε is known as permitivity and is dependent on the medium between the plates. If a capacitor can store 1 *coul* of charges when the voltage across it becomes 1 *volt*, its capacitance is quantified as 1 *Farad* or 1 *F*. One *Farad* of capacitance is too large a quantity to be common. A more practical size for a capacitance is in the order of 10^{-6} *Farad* or 1 micro-Farad, 1 μF. For the current through a capacitor, we have

$$i(t) = \frac{dq}{dt} = C \frac{dv}{dt}$$

In terms of the current i, the voltage v can be represented as follows:

$$v(t) = \frac{1}{C} \int_{-\infty}^{t} i(t)dt = \frac{1}{C} \int_{-\infty}^{0} i(t)dt + \frac{1}{C} \int_{0}^{t} i(t)dt = v(0) + \frac{1}{C} \int_{0}^{t} i(t)dt$$

where $v(0)$ is often referred to as the initial voltage. The symbol for a capacitor resembles that of a couple of parallel plates as shown in **Figure 1.11**.

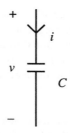

Figure 1.11. The symbol for a capacitor

The energy stored in a capacitor can be derived by performing the integration of the power $p(t)$ as follows:

$$w(t) = \int_{-\infty}^{t} p(t)dt = \int_{-\infty}^{t} v(t)i(t)dt = \int_{-\infty}^{t} v(t) \cdot C \frac{dv}{dt} dt = \int_{-\infty}^{t} v(t) \cdot Cdv$$

$$= \frac{C}{2} v^2(t) \Big|_{-\infty}^{t} = \frac{C}{2} v^2(t) - \frac{C}{2} v^2(-\infty) = \frac{C}{2} v^2(t)$$

In the equation, it is assumed that the voltage across the capacitor is zero at $t = -\infty$. Note that the energy stored in a capacitor is always greater than or equal to zero. This energy is dependent on the square of the voltage across the capacitor and is independent of how the voltage is acquired.

1.4.2 Inductors

When a piece of thin wire is wrapped around a core made of ceramic or iron, an inductor is thus formed. The wire so-formed is called a coil as shown in **Figure 1.12**.

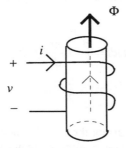

Figure 1.12. A coil formed by wrapping wire around a cylindrical core

14

Within the core, there is strong presence of a magnetic field when the current is flowing in the wire. The strength of the magnetic field represented by the flux Φ is found to be approximately linearly proportional to the current i. When the current is kept constant, the voltage across the coil is zero. However, a voltage is induced inside the coil if the flux is changing. The induced voltage gives rise to the terminal voltage v.

The amount of voltage induced in an inductor is proportional to the time rate of the flux changing. Hence, the amount of voltage induced is proportional to the time rate of the current changing. Thus we have

$$v(t) = L\frac{di}{dt}$$

where the proportional constant L is known as the inductance of the coil. If 1 *volt* of terminal voltage is induced across a coil when the current is changing at a rate of 1 *A* per second, the inductance is quantified as 1 *Henry* or 1 *H*. A common value for an inductance is in the order of 10^{-3} *Henry* or 1 milli-Henry, 1 *mH*. The symbol for an inductor takes after the shape of a coil as shown in **Figure 1.13**.

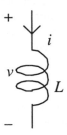

Figure 1.13. The symbol for an inductor

The energy stored in an inductor can be derived by performing the integration of the power $p(t)$ as follows:

$$w(t) = \int_{-\infty}^{t} p(t)dt = \int_{-\infty}^{t} v(t)i(t)dt = \int_{-\infty}^{t} L\frac{di}{dt}i(t)dt = \int_{-\infty}^{t} L \cdot i(t)di$$

$$= \frac{L}{2}i^2(t)\bigg|_{-\infty}^{t} = \frac{L}{2}i^2(t) - \frac{L}{2}i^2(-\infty) = \frac{L}{2}i^2(t)$$

In the equation, it is assumed that the current through the inductor is zero at $t = -\infty$. Note that the energy stored in an inductor is always greater than or equal to zero.

This energy is dependent of the square of the current through the inductor and is independent on how the current is established.

Problems

1.1-1 The current that enters an element is as shown in **Figure 1.14**. Find the charge that enters the element in the time interval $0 < t < 5$ sec.

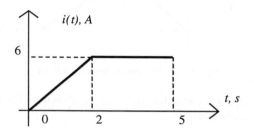

Figure 1.14. The current through an element

1.1-2 Calculate the power absorbed by each of the elements in **Figure 1.15**.

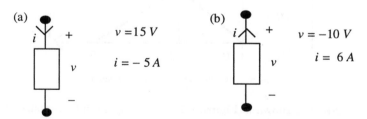

Figure 1.15. Electric elements with different operating conditions

1.1-3 Calculate the power absorbed by each of the elements in **Figure 1.16**.

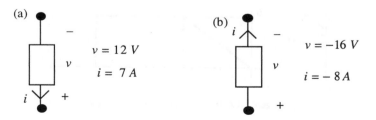

Figure 1.16. Electric elements with different operating conditions

1.3-1. The current in a 5Ω resistor is shown in **Figure 1.17**. Find the power absorbed by the resistor. What is the total energy absorbed by the resistor over the first 8 seconds?

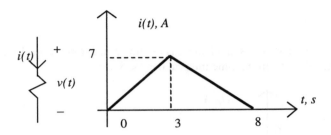

Figure 1.17. The current through a resistor

1.4-1 The voltage across a capacitor of 50 μF is shown in **Figure 1.18**. Find the current through the capacitor.

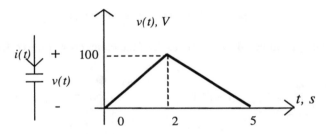

Figure 1.18. The voltage across a capacitor

1.4-2. The current as shown in **Figure 1.19** is flowing through a capacitor of 4 μF. Find the voltage across the capacitor.

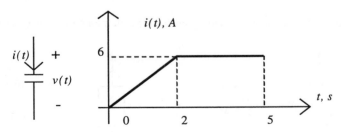

Figure 1.19. The current through a capacitor

1.4-3 The voltage across an inductor of 8 *mH* is shown in **Figure 1.20**. Find the current through the inductor.

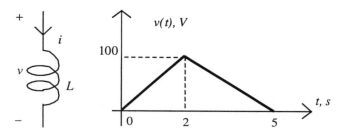

Figure 1.20. The voltage across an inductor

1.4-4. The current as shown in **Figure 1.21** is flowing through a capacitor of 7 *mH*. Find the voltage across the capacitor.

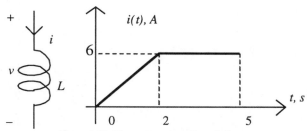

Figure 1.21. The current through an inductor

Chapter 2

Resistive Circuits

A circuit is said to be resistive if it contains only resistors and sources. In such a circuit, no dynamics exists. All the characteristics are describable in terns of algebraic equations. These equations are obtained by applying a couple of circuit laws to two distinctive features in the circuit: a node and a loop. A node is a location at which two or more terminals from different circuit elements are connected together. Two nodes are considered of the same node if there is a piece of wire connecting them directly. A loop is a closed path consisting of two or more circuit elements that are connected terminal to terminal.

2.1 Circuit Laws

There are a couple of fundamental laws concerning circuits in general. These two laws are independent of the elements in the circuits. They are known as circuit laws: the Kirchhoff's Current Law and the Kirchhoff's Voltage Law (sometimes abbreviated as KCL and KVL, respectively). The Kirchhoff's Current Law applies to nodes in circuits while the Kirchhoff's Voltage Law applies to loops in circuits.

2.1.1 Kirchhoff's Current Law

A node can be conveniently considered as a cross-section of a conductor with no volume. Consequently, a node cannot accumulate any charges while charges can neither be created nor be destroyed. This characteristics of a node leads to the Kirchhoff's Current Law which can be stated as follows:

The algebraic sum of currents leaving a node is always zero. When the summation is carried out, a current with direction pointing away from the node is given a positive sign and a current with direction pointing toward the node is assigned a negative sign.

Let's examine the situation in **Figure 2.1**. In **Figure 2.1**, there are five terminals connected together to form a node. Currents i_1, i_2, and i_4 point toward the node while currents i_3 and i_5 point away from the node. Hence, we have

$$-i_1 - i_2 + i_3 - i_4 + i_5 = 0 \qquad\qquad 2.1\text{-}1$$

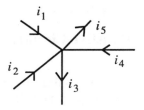

Figure 2.1. A node

Equation 2.1-1can also be rewritten as

$$i_3 + i_5 = i_1 + i_2 + i_4 \qquad\qquad 2.1\text{-}2$$

Consequently, Kirchhoff's Current Law can effectively be stated as follows:

At any node, the sum of all the currents leaving a node equals to the sum of all currents entering the node.

2.1.2 Kirchhoff's Voltage Law

In **Chapter 1**, a voltage between two points is defined as the electrical energy gained by a unit of charges being moved from one point to the other. If this unit of charges is being moved from a particular node of a circuit through a loop and back to the same node, the amount of electrical energy gained by the charges should be zero. If not equal to zero, the conservation principle of energy is violated. Based on this argument, the Kirchhoff's Voltage Law can be stated as follows:

When a loop is traversed in a specific direction, the algebraic sum of voltages encountered is zero. When the algebraic summation is taken, all the voltage drops (from + to −) are assigned one sign (say positive) while all the voltage rises (from − to +) are assigned the opposite sign (say negative).

Let's take a look the situation shown in **Figure 2.2**. In the loop shown, the clockwise direction for traversing is assumed. While traversing the loop, the voltages v_1, v_2, and v_4 are voltage rises encountered while voltage v_3 is a voltage drop encountered. Hence, we have

$$-v_1 - v_2 + v_3 - v_4 = 0 \qquad\qquad 2.1\text{-}3$$

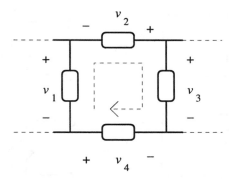

Figure2.2 A loop consists of 4 elements

Equation 2.1-3 can also be rewritten as

$$v_3 = v_1 + v_2 + v_4 \qquad\qquad 2.1\text{-}4$$

Consequently, the Kirchhoff's Voltage Law can effectively be stated as follows:

> When a loop is traversed in a specific direction, the sum of all voltage drops encountered equals to the sum of all voltage rises encountered.

2.2 Equivalent Resistors

Although we can analyze any resistive circuit with the Kirchhoff's Current Law and the Kirchhoff's Voltage Law, there are more efficient methods. One way to minimize the number of equations needed to solve a circuit problem is to recognize a pair of basic connections among resistors: in series or in parallel. When resistors are connected in either fashion, an equivalent circuit may be used to simplify the analysis.

2.2.1 Equivalence

In **Figure 2.3**, there are two boxes. Each box contains only resistors connected in a specific manner. Out of each box is a pair of terminals for external connection. Each box is referred as a one-port (a pair of terminals) network. Two one-port networks are said to be equivalent if they have the same voltage-current characteristics. Since both one-port networks consist of resistors only, they are expected to have the same characteristics as that of a resistor. Consequently, each network is equivalent to a single resistor. If the two equivalent resistors are of the same resistance, the two networks are said to be equivalent.

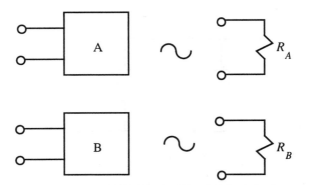

Figure 2.3. Two networks consisting of resistors and their equivalents

2.2.2 Resisters in Series

When a group of resistors are connected end to end as illustrated in **Figure 2.4**, these resistors are said to be in series.

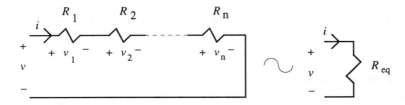

Figure 2.4. Resistors in series and the equivalent resistor

According to the Kirchhoff's Current Law, current i flows through each of the resistors in series. Consequently, the current i causes a voltage drop iR_k across each R_k according to Ohm's Law. By applying the Kirchhoff's Voltage Law, we have

$$v = v_1 + v_2 + \cdots \cdot + v_n = iR_1 + iR_2 + \cdots \cdot + iR_n$$
$$= i(R_1 + R_2 + \cdots \cdot + R_n) = iR_{eq}$$

Hence,

$$R_{eq} = R_1 + R_2 + \cdots \cdot + R_n \qquad 2.2\text{-}1$$

Furthermore, the voltage across each resistor R_k is

$$v_k = i \cdot R_k = \frac{v}{R_1 + R_2 + ... + R_n} \cdot R_k = \frac{R_k}{R_1 + R_2 + ... + R_n} \cdot v \qquad 2.2\text{-}2$$

where k varies from 1 to n.

Equation 2.2-2 suggests that the voltage v is divided into n parts among n resistors. The voltage across a particular resistor is proportional to the resistance of the corresponding resistor when the total resistance is kept constant.

2.2.3 Resisters in Parallel

When a group of resistors are connected between two nodes as shown in **Figure 2.5**, they are said to be in parallel.

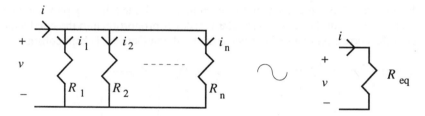

Figure 2.5. Resistors in parallel and the equivalent resistor

Each resistor has the same voltage drop v. This voltage drop produces a current i_k in R_k. By Ohm's Law, we have

$$i_k = \frac{v}{R_k}$$

where k ranges from 1 through n. Furthermore, according to KCL, we have

$$i = i_1 + i_2 + \cdots\cdots + i_n = \frac{v}{R_1} + \frac{v}{R_2} + \cdots\cdots + \frac{v}{R_n}$$

$$= v(\frac{1}{R_1} + \frac{1}{R_2} + \cdots\cdots + \frac{1}{R_n}) = \frac{v}{R_{eq}}$$

Hence,

$$\frac{1}{R_{eq}} = \frac{1}{R_1} + \frac{1}{R_2} + \cdots\cdots + \frac{1}{R_n} \qquad 2.2\text{-}3$$

24

or, equivalently,

$$R_{eq} = (\frac{1}{R_1} + \frac{1}{R_2} + \cdots + \frac{1}{R_n})^{-1}$$ 2.2-4

Furthermore, we have

$$i_k = \frac{v}{R_k} = \frac{i}{\dfrac{1}{R_1} + \dfrac{1}{R_2} + \cdots + \dfrac{1}{R_n}} \cdot \frac{1}{R_k} = \frac{\dfrac{1}{R_k}}{\dfrac{1}{R_1} + \dfrac{1}{R_2} + \cdots + \dfrac{1}{R_n}} \cdot i$$ 2.2-5

where k varies from 1 to n.

Equation 2.2-5 suggests that the current i is divided into n parts among n resistors. The current through a particular resistor is proportional to the conductance $(1/R_k)$ of the corresponding resistor when the total conductance is kept constant.

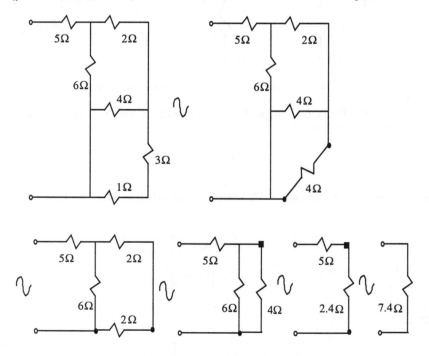

Figure 2.6. A resistive one-port network and its equivalent circuits

2.2.4 Equivalent Resistor via Series and Parallel Combinations

Using series and parallel combinations, we can find the equivalent resistors of a certain kind of one-port network containing only resistors. However, care must be taken to identify whether a group of resistors are connected in series, in parallel, or neither. As a rule of thumb, resistors are in series if they have the same current flowing through each of them and resistors are in parallel if they have the same voltage across each of them. An example of using series and/or parallel combinations for equivalent resistance of an one-port network is illustrated in **Figure 2.6**.

2.3 Node-Voltage Method

A voltage is generally defined between any two points. If a node in a circuit is chosen as the reference, any other node can be assigned a voltage with the understanding that the voltage is defined between this other node and the reference. This common reference node is referred to as the ground. As a rule of thumb, the node with the most elements connected to it is chosen as the ground. The voltage between any two nodes can then be represented as the difference of two node voltages. Referring to **Figure 2.7**, we have

$$v_{ij} = v_i - v_j$$

When one of the node is the ground, the voltage of the ground with respect itself is of course zero.

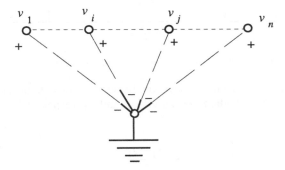

Figure 2.7. Node voltages and the ground

With such an understanding about node voltages, KVL is automatically satisfied. Consequently, only KCL needs to be applied to any node other than the ground for a circuit equation. The number of linearly independent circuit equations is equal to the number of nodes minus one. If a node (not the ground) is connected to other nodes through resistors or current sources, KCL can be applied directly to that node to get a circuit equation linearly independent with the rest. If a node is connected to another node through a voltage source, modification is needed to obtain the correct equation. These two cases are treated separately in the following sections.

2.3.1   Circuits with No Voltage Source

In a circuit containing no voltage source, we get one equation for every node other than the ground by applying KCL to that node. The equation obtained at the ith node takes the following format

$$\sum_{\substack{resistors\cdot connected\cdot \\ to\cdot the\cdot ith\cdot node}} \frac{v_i - v_j}{R_{ij}} + \sum_{\substack{current\cdot sources \\ \cdot leaving\cdot the\cdot node}} I_{slij} = \sum_{\substack{current\cdot sources \\ \cdot entering\cdot the\cdot node}} I_{seij}$$

Let's examine the circuit in **Figure 2.8**.

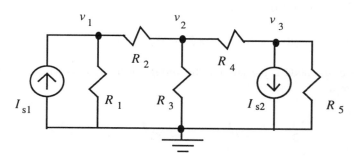

Figure 2.8. A resistive circuit with no voltage source

There are four distinctive nodes in the circuit. One node is chosen as the ground. Other nodes are identified as v_1, v_2, and v_3, respectively. At node v_1, we have

$$\frac{v_1}{R_1} + \frac{v_1 - v_2}{R_2} = I_{s1}$$

At node v_2, we have

$$\frac{v_2 - v_1}{R_2} + \frac{v_2}{R_3} + \frac{v_2 - v_3}{R_4} = 0$$

At node v_3, we have

$$\frac{v_3}{R_5} + \frac{v_3 - v_2}{R_4} + I_{s2} = 0$$

There are three node voltages and three linearly independent equations. Node voltages can be found by solving the simultaneous equations obtained.

2.3.2 Circuits Containing Voltage Sources

A voltage source fixes the voltage between the two nodes where the source is connected. For this pair of nodes illustrated in **Figure 2.9**, one of the two independent equations can easily be identified. Namely,

$$V_s = v_i - v_j \qquad\qquad 2.3\text{-}1$$

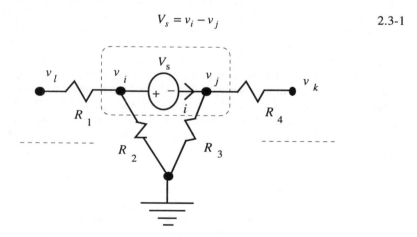

Figure 2.9. A circuit containing a voltage source

If one of the nodes is the ground, **Equation 2.3-1** with one of the node voltage equal to zero is the only independent equation needed. If neither nodes is the ground and there is no other voltage source attached to either node, the second independent equation can be found with KCL. For the time being, the current in the voltage source is identified as i as shown. At node v_i, we have

$$\frac{v_i}{R_2} + \frac{v_i - v_l}{R_1} + i = 0 \qquad\qquad 2.3\text{-}2$$

At node v_j, we have

$$\frac{v_j}{R_3} + \frac{v_j - v_k}{R_4} = i \qquad\qquad 2.3\text{-}3$$

Current i is eliminated when the **Equations 2.3-2** and **2.3-3** are added together. Consequently, we have

$$\frac{v_i}{R_2} + \frac{v_i - v_l}{R_1} + \frac{v_j}{R_3} + \frac{v_j - v_k}{R_4} = 0$$

This equation can actually be found by considering that the two nodes with the voltage source attached together form a 'supernode'. The term supernode does not imply that the two nodes are of the same voltage. As a matter of fact, **Equation 2.3-1** indicates that they are not of the same voltages. Instead, the term supernode implies that we can apply KCL to the supernode directly. Namely, the sum of all the currents entering the supernode equals to the sum of all the currents leaving the supernode. If a group of nodes (the ground excluded) are connected through voltage sources, the node group together with the voltage sources attached can be treated as a supernode. With KCL directly applied to the supernode, the last of the independent equations associated with the node group can be found. However, no other equation is needed if the node group contains the ground. Let's examine the circuit in **Figure 2.10**.

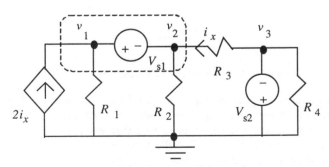

Figure 2.10. A resistive circuit containing a voltage source

There are two voltage sources in the circuit of **Figure 2.10**. Hence, two independent equations concerning these sources can be written down right away. We have

$$V_{s1} = v_1 - v_2$$

and

$$V_{s2} = 0 - v_3 = -v_3$$

The last independent equation is obtained by applying KCL to the supernode formed with nodes v_1 and v_2 plus the voltage source V_{s1}. Hence, we have

$$\frac{v_1}{R_1} + \frac{v_2}{R_2} + \frac{v_2 - v_3}{R_3} = 2i_x$$

Since the current i_x is a variable different from the node voltages assigned, an additional equation for the current is found to facilitate the solving of the node voltages as follows:

$$i_x = \frac{v_3 - v_2}{R_3}$$

2.4 Loop-Current Method

In a circuit carrying some current, there is at least one loop. Starting from any loop, a current circulating that loop can be assigned. A circulating current satisfies KCL automatically. This current is referred to as a loop-current. For any additional loop containing at least one new element that is not in any previous loops with corresponding loop currents already assigned, a new loop current can be identified. Loop-currents are meaningful and useful if the following convention is established:

When a circuit element is contained in only one loop with a corresponding loop-current, the current in the element identified with the same direction as the loop current is equal to the loop-current. When a circuit element is contained in two or more loops, the current in the element is equal to the algebraic sum of all the corresponding loop-currents of the loops that contains the element. The sum is formed by attaching positive signs to those loop-currents flowing in the same direction as that of the element current and by attaching negative signs to those loop-current flowing against the element current.

Loop currents so identified lead us to linearly independent circuit equations when KVL is applied to each of them. The number of linearly independent circuit equations is equal to the number of loop-currents identified. If a loop contains no current sources, KVL can be applied directly to that loop to get a circuit equation linearly independent with the rest. If a loop contains a current source, modification is needed to obtain the correct equation. These two cases are treated separately in the following sections.

2.4.1   Circuits with No Current Source

In a circuit containing no current source, we get one equation for every loop identified with a loop-current. The equation obtained for the *kth* loop takes the following format

$$\underset{\substack{resistors\cdot in\cdot \\ the\cdot kth\cdot loop}}{\sum I_{kj}R_{kj}} + \underset{\substack{voltage\cdot source\cdot drops\cdot in \\ \cdot the\cdot loop-current\cdot direction}}{\sum V_{sdkj}} = \underset{\substack{voltage\cdot source\cdot rises\cdot in \\ \cdot the\cdot loop-current\cdot direction}}{\sum V_{srkj}}$$

where I_{kj} denotes the algebraic sum of loop currents that share the resistor R_{kj}.

Let's examine the circuit in **Figure 2.11**.

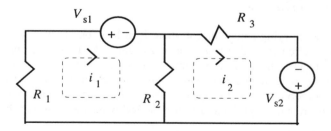

Figure 2.11. A resistive circuit containing no current source

There are two independent loops assigned with corresponding loop-currents. When KVL is applied to the loop identified with i_1, we have

$$i_1 R_1 + (i_1 - i_2)R_2 + V_{s1} = 0$$

When KVL is applied to the loop identified with i_2, we have

$$i_2 R_3 + (i_2 - i_1)R_2 = V_{s2}$$

There are two unknowns and there are two linearly independent circuit equations. Loop-currents i_1 and i_2 can be found by solving the independent equations.

2.4.2 Circuits Containing Current Sources

A current source is identified by a current quantity of its own. If the current source is contained in only one loop assigned with a loop-current, then the current quantity is equal to the loop-current when they have the same direction. Otherwise, the current quantity equals to the loop-current multiplied by -1. If the current source is shared by a set of loops with corresponding loop-currents, the current quantity equals to the algebraic sum of all the corresponding loop-currents of the loop set. This give rise to one independent equation. The rest of the independent equations associated with the loop set are obtained by applying KVL to loops in the set. In **Figure 2.12**, a current source is shared by two loops identified with loop-currents loop i_k and i_j. Note that loops and the corresponding loop currents are identified with dotted lines. One of the two independent equations associated with these two loops can easily be identified. Namely,

$$I_s = i_j - i_k \qquad\qquad 2.4\text{-}1$$

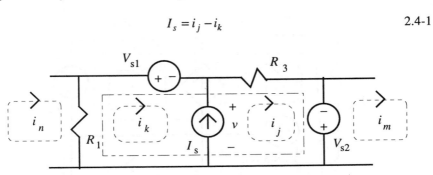

Figure 2.12. A resistive circuit containing a current source

The second independent equation is obtained by applying KVL to the two loops containing the current source. For the time being, let's assign the voltage across the current source to be v. For the loop identified with the loop current i_k, we have

$$(i_k - i_n)R_1 + V_{s1} + v = 0 \qquad\qquad 2.4\text{-}2$$

For the loop identified with the loop current i_j, we have

$$i_j R_3 = V_{s2} + v \qquad\qquad 2.4\text{-}3$$

The voltage v is eliminated when **Equations 2.4-2** and **2.4-3** are added together. Hence, we have

$$(i_k - i_n)R_1 + V_{s1} + i_j R_3 = V_{s2}$$

This equation can actually be found by merging the two loops sharing the current source to form a 'superloop' identified by dot-dash lines. When KVL is applied to the superloop, the correct independent circuit equation is obtained. Note that no new loop-current should be assigned to the superloop and that the current source is not part of the superloop. If a group of loops with corresponding loop-currents are sharing current sources, a superloop that contains no current source may be formed by merging two or more loops. In such a case, the same kind of carefulness as in assigning loop-currents must be taken to avoid forming an unnecessary superloop.

Let's examine the circuit in **Figure 2.13**.

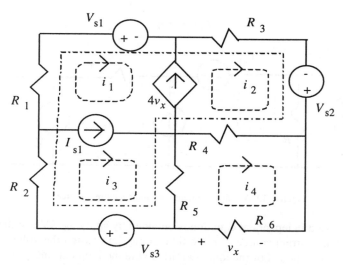

Figure 2.13. A resistive circuit containing current sources

In **Figure 2.13**, four loops with corresponding loop-currents are identified with dotted lines. There are two current sources. Each one gives rise to an independent circuit equation. Consequently, we have

$$I_{s1} = i_3 - i_1$$

and

$$i_2 - i_1 = 2v_x = 4(-i_4 R_6)$$

Note that the voltage v_x is replaced by an expression in terms of the loop currents assigned. Loop i_4 contains no current source. Applying KVL to the loop, we have

$$(i_4 - i_2)R_4 + (i_4 - i_3)R_5 + i_4 R_6 = 0$$

Loops i_1, i_2, and i_3, share two current sources. Merging these loops all together, a superloop identified with dot-dash lines is formed. Applying KVL to this superloop, we have the last independent equation as follows

$$i_1 R_1 + V_{s1} + i_2 R_3 - V_{s2} + (i_2 - i_4)R_4 + (i_3 - i_4)R_5 - V_{s3} + i_3 R_2 = 0$$

2.5 Operational Amplifiers

An operational amplifier (abbreviated as Op-Amp) is an integrated circuit and usually equipped with eight terminals. The detailed characteristics of an Op-Amp is very complicated and beyond the scope of this book. However, Op-Amps are very useful and versatile in many applications and the function of an Op-Amp can be explained with a simple model. In the following sections, some applications of Op-Amps are discussed.

2.5.1 The Ideal Op-Amp

When a Op-Amp is properly connected and adjusted, it can be represented by a model shown in **Figure 2.14**. Note that only 4 terminals are shown: a positive terminal, a negative terminal, an output terminal, and a ground.

For an ideal Op-Amp, the current entering the positive terminal is zero and so is the current entering the negative terminal. Hence, we have

$$i_1 = i_2 = 0$$

Furthermore, the voltage at the positive terminal is equal to the voltage at the negative terminal. Hence, we have

$$v_1 = v_2$$

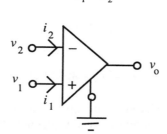

Figure 2.14. The model of an ideal Op-Amp

The output voltage v_o of the ideal Op-Amp is unspecified and is dependent on the connections to the rest of the circuit. Nevertheless, this voltage must remain between $-V_o$ and $+V_o$ with a pre-specified positive voltage V_o for the device to be linear.

2.5.2 The Inverting Op-Amp

A very useful application of an Op-Amp is for signal amplification. One way to achieve signal amplification is demonstrated in **Figure 2.15**.

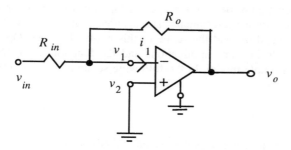

Figure 2.15. An inverting Op-Amp

Applying KCL to the negative terminal of the Op-Amp, we have

$$\frac{v_{in} - v_1}{R_{in}} + \frac{v_o - v_1}{R_o} = i_1$$

Solving for v_o and using the properties of an ideal Op-Amp, we find

$$v_o = -\frac{R_o}{R_{in}}v_{in} + v_1 + \frac{R_o}{R_{in}}v_1 + i_1 R_o = -\frac{R_o}{R_{in}}v_{in}$$

This equation suggests that the circuit amplify the input signal with an amplification value defined as follows:

$$\alpha = \frac{v_o}{v_{in}} = -\frac{R_o}{R_{in}}$$

Due to the negative value of the amplification, this circuit is known as the inverting Op-Amp.

2.5.3 The Non-Inverting Op-Amp

A different way to achieve signal amplification without introducing a negative amplification is demonstrated in **Figure 2.16**.

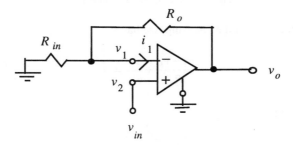

Figure 2.16. A non-inverting Op-Amp

Applying KCL to the negative terminal of the Op-Amp, we have

$$\frac{v_1}{R_{in}} + \frac{v_1 - v_o}{R_o} + i_1 = 0$$

Solving for v_o and using the properties of an ideal Op-Amp, we find

$$v_o = \left(1 + \frac{R_o}{R_{in}}\right)v_1 + i_1 R_o = \left(1 + \frac{R_o}{R_{in}}\right)v_{in}$$

The amplification for the non-inverting Op-Amp is therefore found to be

$$\alpha = \frac{v_o}{v_{in}} = 1 + \frac{R_o}{R_{in}}$$

2.5.4 The Difference Amplifier

When the difference of two given signals is to be examined, the circuit in **Figure 2.17** offers amplification.

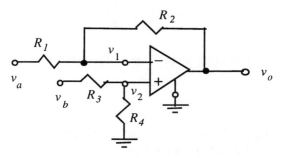

Figure 2.17. A difference amplifier

At the positive terminal of the Op-Amp, we have

$$\frac{v_2 - v_b}{R_3} + \frac{v_2}{R_4} = 0$$

Solving for v_2, we have

$$v_2 = \frac{R_4 v_b}{R_3 + R_4}$$

At the negative terminal of the Op-Amp, we have

$$\frac{v_1 - v_a}{R_1} + \frac{v_1 - v_o}{R_2} = 0$$

Solving for v_o and using the properties of an ideal Op-Amp, we find

$$v_o = \left(1 + \frac{R_2}{R_1}\right) v_1 - \frac{R_2 v_a}{R_1} = \left(1 + \frac{R_2}{R_1}\right) \frac{R_4 v_b}{R_3 + R_4} - \frac{R_2 v_a}{R_1}$$

If the resistors are so chosen that

$$\frac{R_2}{R_1} = \frac{R_4}{R_3} ,$$

the equation for v_o then becomes

$$v_o = \left(1 + \frac{R_4}{R_3}\right)\frac{R_4 v_b}{R_3 + R_4} - \frac{R_2 v_a}{R_1} = \frac{R_4 v_b}{R_3} - \frac{R_2 v_a}{R_1} = \frac{R_2}{R_1}(v_b - v_a)$$

2.5.5 The Summing Amplifier

When a few signals are to be added together, the following circuit performs the summation while offering various summing weights for different signals.

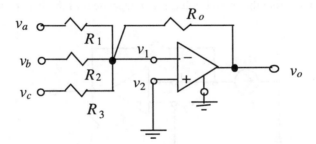

Figure 2.18. A summing amplifier

At the negative terminal, we have

$$\frac{v_1 - v_a}{R_1} + \frac{v_1 - v_b}{R_2} + \frac{v_1 - v_c}{R_3} + \frac{v_1 - v_o}{R_o} = 0$$

Solving for v_o and using the properties of an ideal Op-Amp, we find

$$v_o = \left(1 + \frac{R_o}{R_1} + \frac{R_o}{R_2} + \frac{R_o}{R_3}\right)v_1 - \frac{R_o v_a}{R_1} - \frac{R_o v_b}{R_2} - \frac{R_o v_c}{R_3}$$

$$= -\left(\frac{R_o v_a}{R_1} + \frac{R_o v_b}{R_2} + \frac{R_o v_c}{R_3}\right)$$

2.5.6 The Current to Voltage Converter

When a current signal is to be converted into a voltage signal, the circuit in **Figure 2.19** performs the conversion.

At the negative terminal, we have

$$I_{in} + \frac{v_1 - v_o}{R} = 0$$

Solving for v_o and using the properties of an ideal Op-Amp, we find

$$v_o = v_1 + RI_{in} = RI_{in}$$

It should be noted that the output voltage is independent of the load resistance R_L.

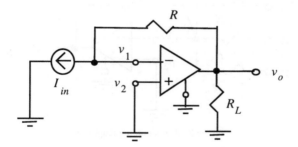

Figure 2.19. A current to voltage converter

Problems

2.1-1. Find the current i_x and the voltage v of the circuit shown in **Figure 2.20**.

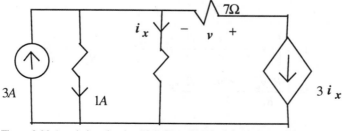

Figure 2.20 A resistive circuit with independent and dependent current sources

2.1-2. Find the current i_x of the circuit shown in **Figure 2.21**.

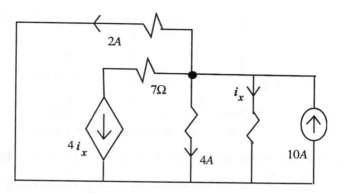

Figure 2.21 A resistive circuit with independent and dependent current sources

2.2-1. Find the equivalent resistance R_{ab} for the circuits shown in **Figure 2.22**.

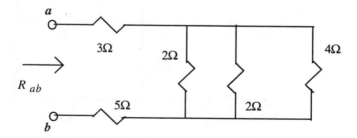

Figure 2.22 A resistive network containing no source

2.2-2. Find the equivalent resistance R_{ab} for the circuits shown in **Figure 2.23**.

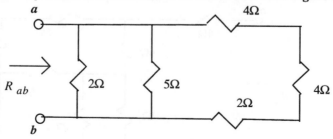

Figure 2.23 A resistive network containing no source

2.2-3. Find the equivalent resistance R_{ab} for the circuits shown in **Figure 2.24**.

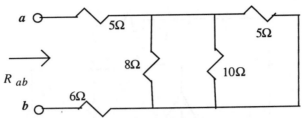

Figure 2.24 A resistive network containing no source

2.2-4. Find the equivalent resistance R_{ab} for the circuits shown in **Figure 2.25**.

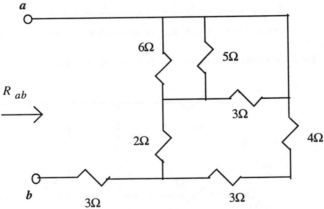

Figure 2.25 A resistive network containing no source

2.2-5. Find the equivalent resistance R_{ab} for the circuits shown in **Figure 2.26**.

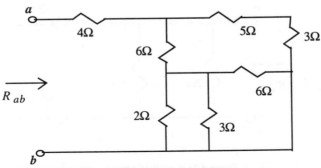

Figure 2.26 A resistive network containing no source

2.2-6. Find the equivalent resistance R_{ab} for the circuits shown in **Figure 2.27**.

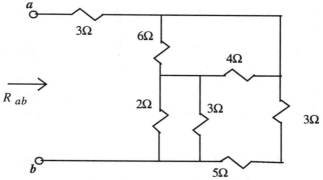

Figure 2.27 A resistive network containing no source

2.3-1. Using the nodal analysis, find the current v_x of the circuit shown in **Figure 2.28**.

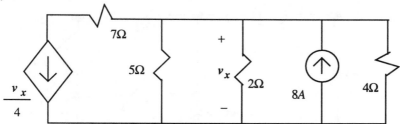

Figure 2.28 A resistive network containing dependent and independent current sources

2.3-2. Using the nodal analysis, find the current i_x of the circuit shown in **Figure 2.29**.

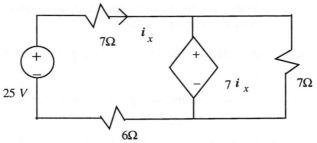

Figure 2.29 A resistive network containing a dependent voltage source

2.3-3. Using the nodal analysis, find the current I_o in the circuit shown **Figure 2.30**.

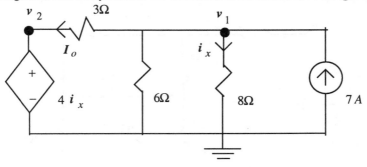

Figure 2.30 A resistive circuit containing a dependent voltage source

2.3-4. Using the node voltages assigned in **Figure 2.31**, write down the network equations necessary for solving these node voltages.

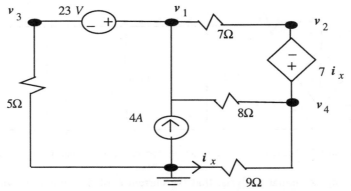

Figure 2.31 A resistive circuit containing independent and dependent sources

2.3-5. Using the node voltages assigned in **Figure 2.32**, write down the network equations necessary for solving these node voltages.

2.4-1. Using the loop analysis, find the current i_x of the circuit shown in **Figure 2.33**.

2.4-2. Using the loop analysis, find the voltage v_x of the circuit shown in **Figure 2.34**.

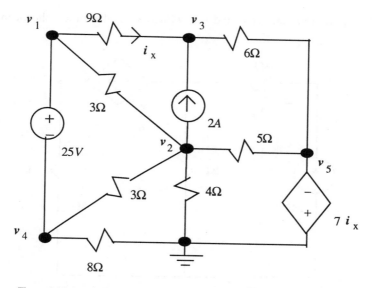

Figure 2.32 A resistive circuit containing independent and dependent sources

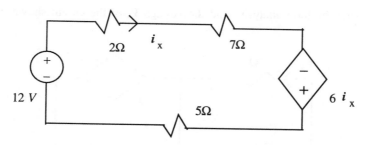

Figure 2.33 A resistive circuit containing independent and dependent voltage sources

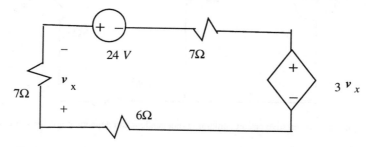

Figure 2.34 A resistive circuit containing independent and dependent voltage sources

2.4-3. Using the loop analysis, find the voltage V_o of the circuit shown in **Figure 2.35**.

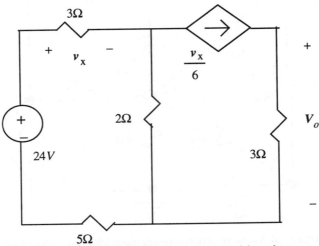

Figure 2.35 A resistive circuit containing independent and dependent sources

2.4-4. Using the loop analysis, find the voltage V_o of the circuit shown in **Figure 2.36**.

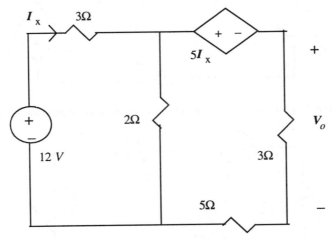

Figure 2.36 A resistive circuit containing independent and dependent voltage sources

2.4-5. Using the loop analysis, find the current I of the circuit shown in **Figure 2.37**.

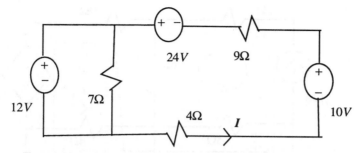

Figure 2.37 A resistive circuit containing independent voltage sources

2.4-6. Using the loop currents assigned in **Figure 2.38,** write down the network equations which are necessary for solving these loop currents.

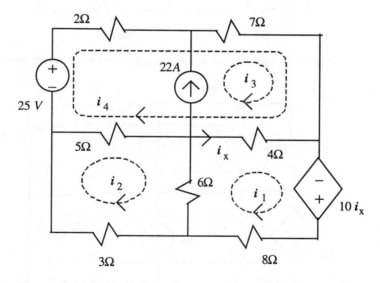

Figure 2.38 A resistive circuit containing independent and dependent sources

2.4-7. Using the loop currents assigned in **Figure 2.39,** write down the network equations which are necessary for solving these loop currents.

2.4-8. Using the loop currents assigned in **Figure 2.40,** write down the network equations necessary for solving these loop currents.

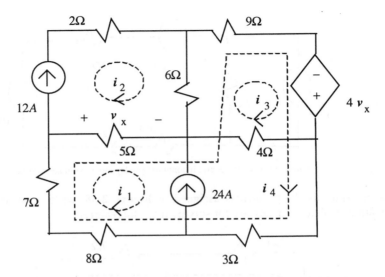

Figure 2.39 A resistive circuit containing independent and dependent sources

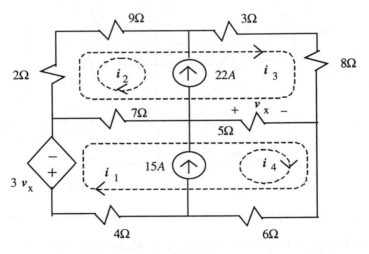

Figure 2.40 A resistive circuit containing independent and dependent sources

2.5-1. For the circuit shown in **Figure 2.41**, find V_o in terms of V_1 and V_2.

2.5-2. For the circuit shown **Figure 2.42**, find V_o in terms of V_1 and V_2.

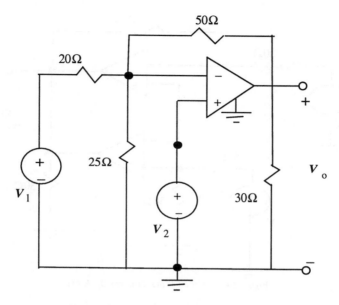

Figure 2.41 A circuit containing an Op-Amp

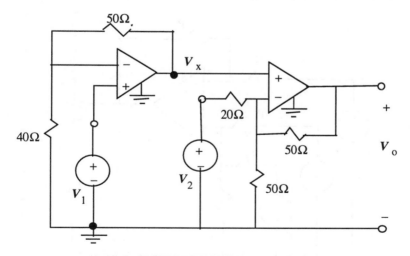

Figure 2.42 A circuit containing two Op-Amps

2.5-3. For the circuit shown in Figure **2.43**, derive the output voltage V_o in terms of source voltages V_a and V_b.

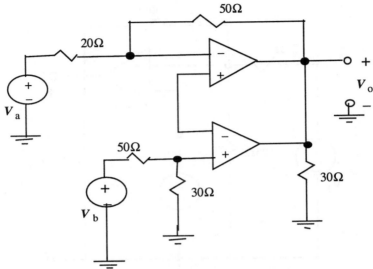

Figure 2.43 A circuit containing two Op-Amps

2.5-4. For the circuit shown in **Figure 2.44**, find the voltage V_o in terms of I_o, V_1 and V_2.

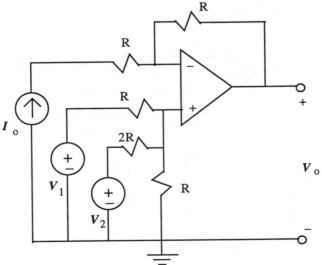

Figure 2.44 A circuit containing an Op-Amp

Chapter 3
Circuit Theorems

Circuit theorems are derived from circuit laws. These theorems often enable us to analyze circuits more efficiently. Well-known circuit theorems are linearity (superposition), Thevenin's Theorem, Norton's Theorem, and Maximum Power Transfer.

3.1 Linearity and Superposition

Resistors are governed by Ohm's Law, which can be represented by a linear function. Resistive circuits that are formed with resistors and sources are known as linear circuits. The current through a resistor or the voltage across a resistor can be represented by a linear function. Using the properties of a linear function, circuit analysis of certain kinds of circuits is made easier.

3.1.1 Linearity

Linearity is a property associated with an operator (also known as function or mapping). Let $x(t)$ and $y(t)$ be any function (generally vector-valued) of time. An operator T that transforms a time function to another is said to be linear if it possesses the following properties

$$\begin{cases} T\{x(t) + y(t)\} = T\{x(t)\} + T\{y(t)\} \\ \quad T\{ax(t)\} = aT\{x(t)\} \end{cases} \qquad 3.1\text{-}1$$

where a is any scalar.

In a resistive circuit, the time function to be transformed consists of independent source functions. Any other circuit quantity can be considered as the output of a linear operator transforming the source functions. This property is very useful in analyzing the kind of circuits known as the ladder networks as shown in **Figure 3.1**.

Example 3.1 For the circuit shown in **Figure 3.1**, V_s *is of 7* V. *Find* v_o.

The voltage v_o can be written as a function of V_s, say $f(V_s)$. Furthermore, we have

$$v_o = f(V_s) = K \times V_s \qquad 3.1\text{-}2$$

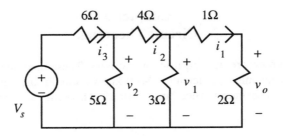

Figure 3.1. A ladder network

The proportional constant K in **Equation 3.1-2** can be determined firstly by assigning v_o a convenient value, say 2 V. The corresponding value for V_s can then be found by working from the right side of the circuit to the left as follows:

$$i_1 = v_o / 2 = 2 / 2 = 1\,A$$

$$v_1 = 1 \times i_1 + v_o = 1 + 2 = 3\,V$$

$$i_2 = i_1 + v_1 / 3 = 1 + 3 / 3 = 2\,A$$

$$v_2 = 4 \times i_2 + v_1 = 8 + 3 = 11\,V$$

$$i_3 = i_2 + v_2 / 5 = 2 + 11 / 5 = 4.2\,A$$

$$V_s = 6 \times i_3 + v_2 = 25.2 + 11 = 36.2\,V$$

The proportional constant K is thus found to be

$$K = v_o / V_s = 2 / 36.2$$

Consequently, the true value of v_o is

$$v_o = K \times V_s = (2 / 36.2) \times 7 = 0.387\,V$$

3.1.2. Superposition

When there are two or more independent sources in a circuit, linearity manifests itself in a special way. Any other circuit quantity can be treated as a function of the vector consisting of all independent source functions. Assuming these source functions are x_1, x_2, ..., through x_n, representing source voltages or source currents. We have

$$[x_1, x_2,..., x_n] = [x_1,0,...,0] + [0, x_2,0,...,0] + ... + [0,0,...,0, x_n]$$

For a linear operator $\mathbf{T}$ operating on the vector, we have

$$\mathbf{T}[x_1, x_2,...,x_n] = \mathbf{T}[x_1,0,...,0] + \mathbf{T}[0, x_2,0,...,0] + ... + \mathbf{T}[0,0,...,0,x_n]$$

This means that we can analyze the circuit under the influence of one independent source at a time by setting the rest of the independent sources to zero. Possible benefits are derived from setting independent sources to zero. For a voltage source, setting the source function to zero means the voltage across the source is zero while the current through the source can take on any value dictated only by the rest of the circuit. Hence, it behaves like a short circuit as shown in **Figure 3.2**.

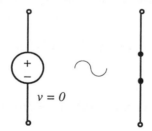

$v = 0$

Figure 3.2. The equivalent circuit of an independent voltage source set to zero

For a current source, setting the source function to zero means the current through the source is zero while the voltage across the source can take on any value dictated only by the rest of the circuit. Hence, it behaves like an open circuit as shown in **Figure 3.3**.

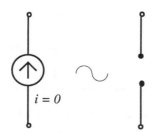

$i = 0$

Figure 3.3. The equivalent circuit of an independent current source set to zero

Let us study the circuit in the following example.

Example 3.2 For the circuit in **Figure 3.4**, V_s is of 10 V and I_s is of 5 A. Find v_o.

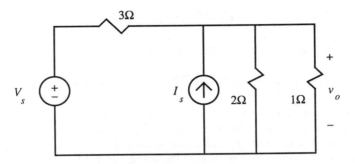

Figure 3.4 A circuit with two independent sources

There are two sources in the circuit. Hence, v_o is a function of a vector of two components. We have

$$v_o = f([V_s, I_s]) = f([V_s, 0]) + f([0, I_s]) = v_{ov} + v_{oi}$$

where

$$v_{ov} = f([V_s, 0]), \quad \text{and} \quad v_{oi} = f([0, I_s])$$

For v_{ov}, we set I_s to zero. The circuit in **Figure 3.4** becomes the one in **Figure 3.5**.

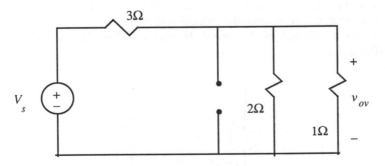

Figure 3.5. The equivalent circuit with the current source set to zero

From the equivalent circuit in **Figure 3.5**, we get

$$v_{ov} = \frac{\dfrac{2 \times 1}{2+1}}{3 + \dfrac{2 \times 1}{2+1}} V_s = \frac{2}{9+2} \times 10 = \frac{20}{11} \; V$$

For v_{ov}, we set I_s to zero. The circuit in **Figure 3.4** becomes the one in **Figure 3.6**.

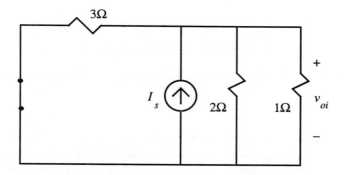

Figure 3.6. The equivalent circuit with voltage source set to zero

From the equivalent circuit in **Figure 3.6**, we get

$$v_{oi} = \frac{1}{\dfrac{1}{3} + \dfrac{1}{2} + \dfrac{1}{1}} I_s = \frac{6}{11} \times 5 = \frac{30}{11} \; V$$

Consequently,

$$v_o = v_{ov} + v_{oi} = \frac{20}{11} + \frac{30}{11} = \frac{50}{11} \; V$$

We can get the very same answer by analyzing the circuit in **Figure 3.4** with the node-voltage method. It should be noted that the principle of superposition may be too time-consuming to apply when there are too many independent sources in a circuit.

3.2 Thevenin's Theorem

The one-port network shown in **Figure 3.7** is among the simplest resistive one-port networks that contain both independent voltage sources and resistors.

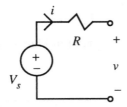

Figure 3.7. The simplest resistive one-port containing a voltage source

The characteristics of such a network is governed by the following equations:

$$v = V_s - i \cdot R$$

The terminal voltage v is equal to V_s if the current i is equal to zero, i.e., if the terminals are left open-circuited. On the other hand, the network is equivalent to a simple resistor with resistance R if V_s is set to zero. Such a characteristic is generally found for an one-port network containing only resistors and sources according to Thevenin's Theorem. The theorem states that any one-port containing only resistors and sources is equivalent to a simple one-port consisting of a voltage source and a resistor in series as illustrated in **Figure 3.8**. In the equivalent circuit, the voltage source V_{Th} is the open-circuited voltage across the two terminals. The resistance R_{Th} is the equivalent resistance of the one-port with all internal independent sources set to zero.

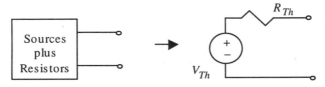

Figure 3.8 Thevenin's equivalent circuit

The usefulness of Thevenin's Theorem lies in the fact that changes may be made to the connections to the terminals from the outside of the one-port network. The effect of such a change can be fully accounted for with the simpler one-port instead of the original more complicated one. Let's examine the following **Example 3.3**.

Example 3.3 In the circuit shown in **Figure 3.9**, the resistor R_L can be either 7Ω or 12Ω. Find the corresponding value for v_o for each case.

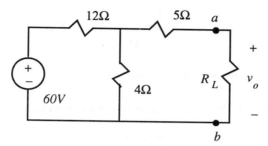

Figure 3.9. A circuit with a variable resistor

Using Thevenin's Theorem, the one-port network that is left to the resistor R_L is equivalent to a simpler one as shown in **Figure 3.8**. For the Thevenin's equivalent voltage source V_{Th}, terminals a and b are left open and the one-port network becomes the one shown in **Figure 3.10**.

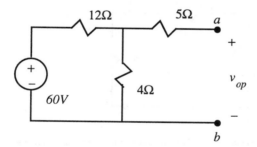

Figure 3.10. The one-port in Figure 3.9 with the port open-circuited

From the circuit in **Figure 3.10**, we have

$$V_{Th} = v_{op} = 60 \times \frac{4}{12+4} = 15 \ V$$

Note that the open-circuit voltage is the same as the voltage across the 4-Ω resistor since there is zero *volt* across the 5-Ω resistor due to the open circuit.

For the Thevenin's equivalent resistance R_{Th}, we set all independent sources in the one-port to zero and the one-port becomes the circuit in **Figure 3.11**.

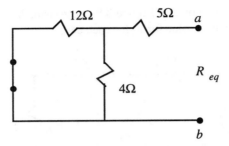

Figure 3.11. The one-port in Figure 3.9 with sources set to zero

From the circuit in **Figure 3.11**, we have

$$R_{Th} = R_{eq} = 5 + \frac{4 \times 12}{4 + 12} = 8\Omega$$

The circuit in **Figure 3.9** then becomes the one shown in **Figure 3.12**.

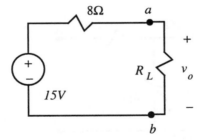

Figure 3.12. The Thevenin's equivalent circuit of the circuit in Figure 3.9

From the circuit in **Figure 3.12**, we have

$$v_o = 15 \times \frac{R_L}{8 + R_L} = \begin{cases} 7V, & if \quad R_L = 7\Omega \\ \\ 9V, & if \quad R_L = 12\Omega \end{cases}$$

3.3 Norton's Theorem

The one-port network shown in **Figure 3.13** is among the simplest resistive one-port networks which contain both independent current sources and resistors.

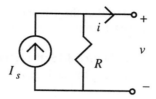

Figure 3.13. The simpliest resistive one-port containing a current source

The characteristics of such a network is governed by the following equations:

$$i = I_s - \frac{v}{R}$$

The current i is equal to I_s if the terminal voltage v is equal to zero, i.e., if the terminals are short-circuited. On the other hand, the network is equivalent to a simple resistor with resistance R if I_s is set to zero. Such a characteristic is demonstrated in any one-port network containing only resistors and sources according to Norton's Theorem. The theorem states that any one-port containing only resistors and sources is equivalent to a simple one-port consisting of a current source and a resistor in parallel as demonstrated in **Figure 3.14**.

Figure 3.14. Norton's equivalent circuit

In the equivalent circuit, the resistance R_{Th} is the same equivalent resistance of the one-port with all the independent sources set to zero as stated in the Thevenin's Theorem. The current source I_{Sh} is the would-be current flowing from one terminal to the other if they were short-circuited. The direction of the current source is such that it would give rise to the same current if the two terminals were shorted together. These parameters are related to the Thevenin's equivalent voltage V_{Th} through the following equation

$$V_{Th} = I_{Sh} \cdot R_{Th}$$

The usefulness of Norton's Theorem is similar to that of Thevenin's Theorem. To effectively account for any change made outside the one-port network, the

58

simpler and equivalent one-port instead of the original more complicated one can be utilized. Let's examine the following example.

Example 3.4 In the circuit shown in **Figure 3.15**, the resistor R_L can be either 4Ω or 12Ω. Find the corresponding value for i_o for each case.

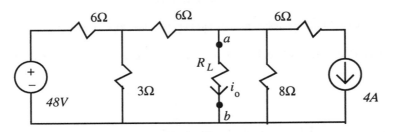

Figure 3.15. A circuit with a variable resistor

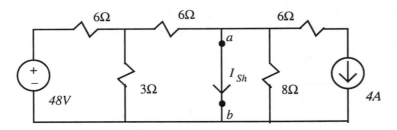

Figure 3.16. The one-port in Figure 3.15 with the port short-circuited

Using Norton's Theorem, the one-port network that is connected to the resistor R_L at terminal a-b is equivalent to a simpler one as shown in **Figure 3.14**. For the Norton's equivalent current source I_{Sh}, terminal a-b are shorted together and the one-port network becomes the one shown in **Figure 3.16**.

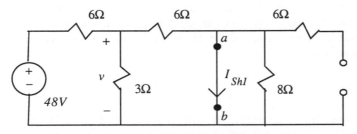

Figure 3.17 The short-circuit current due to the voltage source

Since there are two sources, I_{Sh} is consisted of two parts: I_{Sh1} due to the voltage source and I_{Sh2} due to the current source. To find I_{Sh1}, we set the current source to zero and we have the circuit shown in **Figure 3.17**.

Analyzing the circuit in **Figure 3.17**, we have

$$I_{Sh1} = \frac{v}{6} = \frac{1}{6} \times \frac{\dfrac{3 \times 6}{3+6}}{6 + \dfrac{3 \times 6}{3+6}} \times 48 = 2\,A$$

To find I_{Sh2}, we set the voltage source to zero and we have the circuit shown in **Figure 3.18**.

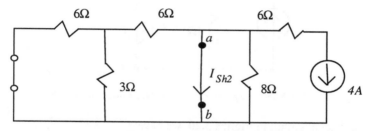

Figure 3.18 The one-port in Figure 3.15 with the port short-circuited

Analyzing the circuit in **Figure 3.18**, we have

$$I_{Sh2} = -4A$$

Therefore,

$$I_{Sh} = I_{Sh1} + I_{Sh2} = 2 - 4 = -2A$$

For R_{Th}, we set all the independent sources to zero and we have the circuit as shown in **Figure 3.19**. Analyzing the circuit in **Figure 3.19**, we have

$$R_{Th} = R_{eq} = \frac{1}{\dfrac{1}{8} + \dfrac{1}{6 + \dfrac{3 \times 6}{3+6}}} = 4\Omega$$

60

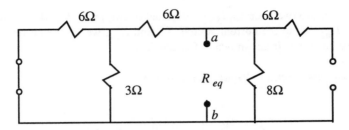

Figure 3.19 Thevenin's equivalent resistance for the one-port in Figure 3.18

The circuit in **Figure 3.15** then becomes the one shown in **Figure 3.20**.

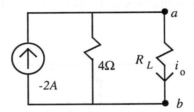

Figure 3.20. The Norton's equivalent circuit of the circuit in Figure 3.15

From the equivalent circuit in **Figure 3.20**, we have

$$i_o = (-2) \times \frac{4}{4+R_L} = \begin{cases} -1A, & \text{if} \quad R_L = 4\Omega \\ \\ -0.5A, & \text{if} \quad R_L = 12\Omega \end{cases}$$

In **Examples 3.3** and **3.4**, the circuits contain no dependent sources. Consequently, the resistances of the Thevenin equivalent resistors were found through series-parallel combinations. For a circuit containing dependent source(s), the method of series-parallel combinations is not applicable. An alternative using Ohm's law is needed. To illustrate this alternative, let's examine **Example 3.5**.

Example 3.5 Find the Thevenin equivalent circuit and the Norton equivalent circuit of the circuit shown in **Figure 3.21**.

Solution: For the Thevenin's equivalent voltage source V_{Th}, we need to find the open-circuit voltage between terminals a and b. Analyzing the circuit in **Figure 3.21**, we have

$$V_{Th} = V_{ab} = 6 \times (9 - i_x) = 7i_x - 5 \times (9 - i_x) - 3i_x$$

Solving for i_x, we have

$$(6+5) \times 9 = (7+5-3+6)i_x \quad i.e. \quad i_x = 99/15 = 6.6A$$

Hence, we have

$$V_{Th} = 6 \times (9 - i_x) = 6 \times (9 - 6.6) = 14.4V$$

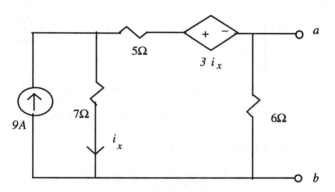

Figure 3.21 A circuit with dependent source

For the Thevenin's equivalent resistance R_{Th}, we set all independent sources in the one-port to zero. The one-port becomes the circuit in **Figure 3.22**.

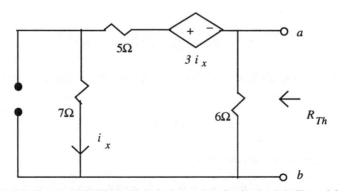

Figure 3.22 The circuit for Thevenin equivalent resistor for the circuit in Figure 3.21

Since the one-port in **Figure 3.22** is equivalent to a resistor, the one-port is governed by the Ohm's Law. Consequently, a *one-A* current source may be connected to the terminals *a-b* to facilitate the finding of the equivalent resistance as shown in **Figure 3.23**.

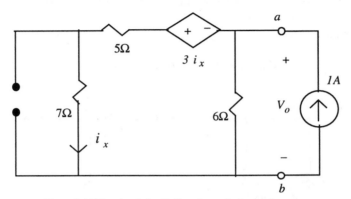

Figure 3.23 The circuit for finding the equivalent resistance

Analyzing the circuit in **Figure 3.23**, we have

$$V_o = 7i_x + 5i_x - 3i_x = 6 \times (1 - i_x)$$

Solving for i_x, we have

$$(7 + 5 - 3 + 6)i_x = 6 \quad i.e. \quad i_x = 6/15 = 0.4A$$

Hence,

$$V_o = (7 + 5 - 3)i_x = 9 \times 0.4 = 3.6V$$

According to the Ohm's Law, the Thevenin equivalent resistance is hence found to be

$$R_{Th} = V_o / 1 = 3.6 / 1 = 3.6\Omega$$

The Norton equivalent current source is therefore

$$I_{Sh} = V_{Th} / R_{Th} = 14.4 / 3.6 = 4A$$

3.4 Maximum Power Transfer

To make the most of a one-port network containing sources, a variable load resistor is usually connected to the one-port. By adjusting the resistance of the load resistor, the power absorbed by the load resistor can be made maximum. With the help of Thevenin's theory, the one-port with the attached load resistor is equivalent to the circuit as shown in **Figure 3.21**.

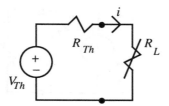

Figure 3.21. The loading of an one-port with a variable resistor

The current i in the load resistor is given as

$$i = \frac{V_{Th}}{R_{Th} + R_L}$$

while the power P_L consumed by the load resistor is found to be

$$P_L = i^2 \cdot R_L = \left(\frac{V_{Th}}{R_{Th} + R_L}\right)^2 R_L$$

The maximum value of P_L is found by taking the derivative of P_L with respective to R_L and set the derivative to zero as follows

$$\frac{dP_L}{dt} = \left(\frac{V_{Th}}{R_{Th} + R_L}\right)^2 - 2\frac{V_{Th}^2 \cdot R_L}{(R_{Th} + R_L)^3} = \frac{V_{Th}^2(R_{Th} + R_L - 2R_L)}{(R_{Th} + R_L)^3} = 0$$

The maximum value of P_L occurs when R_L is adjusted to R_{Th}. Let's study the following example.

Example 3.5 Find the maximum power that can be transferred to the load resistor in the circuit shown in **Figure 3.22**.

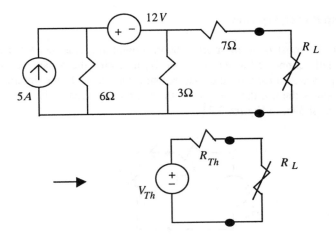

Figure 3.22. A one-port loaded with a variable resistor and its equivalent circuit

For R_{Th}, we examine the following circuit that is obtained from **Figure 3.22** by setting all the independent sources to zero.

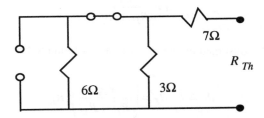

Figure 3.23. The Thevenin's equivalent resistance of the one-port in Figure 3.22

$$R_{Th} = 7 + \frac{6 \times 3}{6 + 3} = 9\Omega$$

For V_{Th}, we use the circuit in **Figure 3.24**.

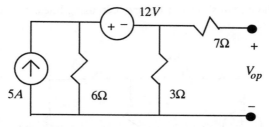

Figure 3.24. The circuit for finding the Thevenin's equivalent voltage
source of the one-port in Figure 3.22

We have

$$\frac{V_{op}}{3} + \frac{V_{op} + 12}{6} = 5$$

Solve for V_{op}, we get

$$V_{op} = \frac{5-2}{\frac{1}{3} + \frac{1}{6}} = 6V = V_{Th}$$

For the maximum power transfer, R_L is adjusted to 9Ω and the maximum power is found to be

$$P_{L,max} = i^2 \cdot R_L = \left(\frac{6}{9+9}\right)^2 \cdot 9 = 1W$$

Problems

3.1-1. Find V_o in the circuit shown in **Figure 3.25** when $I_s = 8A$. Use linearity and start with the assumption that V_o is $2V$.

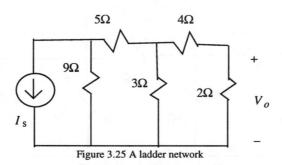

Figure 3.25 A ladder network

3.1-2. Find I_o in the circuit shown in **Figure 3.26** using linearity and the assumption that I_o is $1A$.

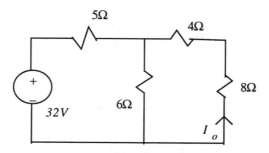

Figure 3.26 A ladder network

3.1-3. Use superposition to find the voltage V_o for the circuit shown in **Figure 3.27**.

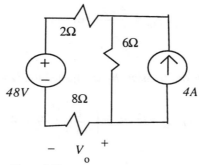

Figure 3.27 A network with two sources

3.1-4. Use superposition to find V_o for the circuit shown in **Figure 3.28**.

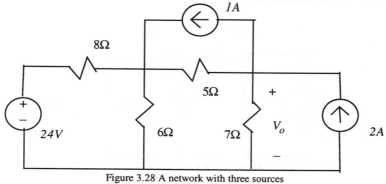

Figure 3.28 A network with three sources

3.2-1. Find the Thevenin equivalent circuit of the network shown in **Figure 3.29**.

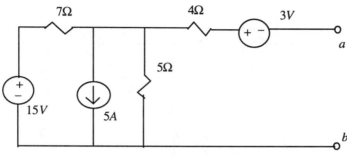

Figure 3.29 A one-port network

3.3-1. Find the Norton equivalent circuit of the network shown in **Figure 3.30**.

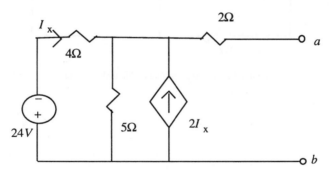

Figure 3.30 A one-port network

3.4-1. Use the Thevenin's Theorem to find the load resistance R_L in **Figure 3.31** for maximum power transfer. Calculate the maximum power transferred.

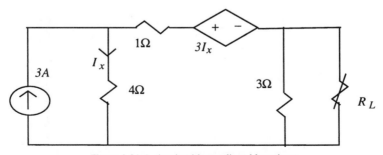

Figure 3.31 A circuit with an adjustable resistor

3.4-2. Use the Norton's Theorem to find the load resistance R_L in **Figure 3.32** for maximum power transfer. Calculate the maximum power transferred.

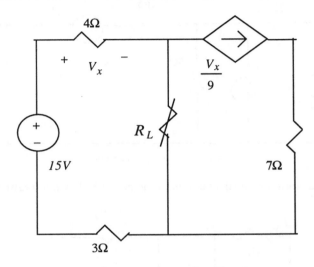

Figure 3.32 A circuit with an adjustable resistor

Chapter 4
Network Equations through Topology

Network equations are obtained with Kirchhoff's Voltage Law and Kirchhoff's Current Law. KVL applies to a loop and KCL applies to a node. Nodes and loops are formed independent of the types of elements in a circuit. Hence, network equations can be formulated based on the information of nodes and loops. This information can be summarized in a simple graph consisting of two types of elements: nodes and branches. A node in a graph corresponds to a node in a circuit while a branch represents an element connected between two nodes. Network topology is a study of the circuit graphs that leads to systematic approaches for obtaining network equations.

4.1 Graph

When a circuit is given, the corresponding graph can be constructed as follows:

1) Each distinctive node in a circuit is identified by a corresponding node in the graph.
2) Each element with its two terminals, independent of types, is identified by a corresponding branch that is connected between two nodes.

Let's examine the following example.

Example 4.1

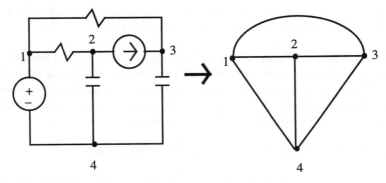

Figure 4.1. A circuit and its corresponding graph

A graph is said to be connected if every node is tied to every other node via some path consisting of branches. In the rest of the chapter, we assume that the graph to be examined is connected.

4.1.1 Tree, Co-tree, and Links

In a graph, there is an important sub-structure called a tree. A tree consists of branches that connect every node to every other node via some path without forming a closed path (i.e., a loop). The easiest way to identify a tree is to start at any node and connect it through a branch to a different node. Subsequently, a new node is added through a new branch to the nodes that are already connected together. Following this approach, there are a total of $N - 1$ branches in a tree if N represents the total number of nodes. Consider the graph in **Figure 4.2**.

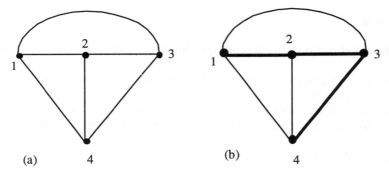

Figure 4.2. A graph with a tree identified and the co-tree

In the graph, there are four nodes. Hence, three branches are needed to form a tree. The three branches shown in (b) with bold-face identify a tree.

The complement of a tree in a graph is called the co-tree. The branches in the co-tree are referred to as links. If B represents the total number of branches in a graph, there are $B - (N - 1)$ links. For the graph with the tree shown in **Figure 4.2**, the three branches identified by hairlines in (b) form the co-tree. The co-tree is not necessary a tree. Let's examine the following example.

Example 4.2 Identify a tree for the graph shown in **Figure 4.3**.

In the graph of **Figure 4.3**, there are five nodes and nine branches. Therefore, a tree contains four branches while its co-tree contains five links. A tree is identified in bold-face as shown in (b). The co-tree is denoted with five hairline branches.

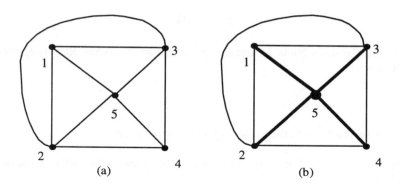

Figure 4.3. A graph with a tree identified and the co-tree

4.1.2 Cut-Sets and Cut-Set Equations

For the graph shown in **Figure 4.2**, let's further identify each branch with an alphabet and an arbitrarily chosen current direction as shown in the **Figure 4.4**.

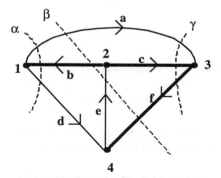

Figure 4.4. A graph with various cut-sets

Also shown in the graph are three different cuts. Each cut is made to sever one node or more from the rest of the graph. Each cut can be identified by a set, known as the cut-set, consisting of the branches severed. For example, cut α severs branches **a**, **b** and **d**. Hence, cut-set α is identified as follows:

$$\alpha = \{a, b, d\}$$

Similarly, we have

$$\beta = \{a, b, e, f\}$$

$$\gamma = \{a, c, f\}$$

Cut-set α isolates Node **1** from the rest of the graph. When KCL is applied to Node **1**, we get

$$i_a + i_d = i_b$$

This equation is known as the cut-set equation for cut-set α.

Cut-set γ isolates Node **3** from the rest of the graph. When KCL is applied to Node **3**, we get

$$i_a + i_c = i_f$$

For cut-set β, Nodes **1** and **4** are isolated from the rest of the graph. When KCL is applied respectively to Node **1** and Node **4**, we have the following equations.

$$i_a + i_d = i_b$$

$$i_e = i_d + i_f$$

Combining the two equations to eliminate i_d, we have

$$i_a + i_e = i_b + i_f$$

This equation is the cut-set equation for the cut-set β. Note that the same equation can be obtained by applying KCL to the supernode formed by Nodes **1** and **4**.

A cut-set that contains exactly one tree branch plus some links is called a *fundamental cut-set*. Since there are $N-1$ tree branches, there are $N-1$ distinctive fundamental cut-sets that lead to $N-1$ independent cut-set equations. For the graph in **Figure 4.4**, the three fundamental cut-sets are

$$\{a, b, d\}, \quad \{d, e, f\}, \quad \{a, c, d, e\}$$

The corresponding cut-set equations are as follows

$$i_a + i_d = i_b$$

$$i_f + i_d = i_e$$

$$i_a + i_c + i_d = i_e$$

4.1.3 Fundamental Loops

A graph representing a circuit normally contains at least one loop. When a tree is identified in a graph, a loop that contains one and only one link is referred to as a *fundamental loop*. When KVL is applied to a fundamental loop, a loop equation is resulted. Since there are $B-N+1$ links, there are $B-N+1$ distinctive fundamental loops that lead to $B-N+1$ independent loop equations.

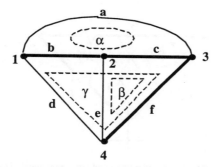

Figure 4.5. A graph with a tree and the fundamental loops

In the graph shown in **Figure 4.5**, branches **a**, **d**, and **e** are links. Three fundamental loops are identified by α, β, and γ. Loop α consists of branches **a**, **b**, and **c**. Loop β consists of branches **c**, **e**, and **f**. Loop γ consists of branches **b**, **c**, **d**, and **f**.

4.2 General Nodal Analysis

With the help of network topology, nodal analysis can be structured in such a way that is independent of whether voltage sources are present. The utilization of fundamental cut-sets makes the forming of supernode unnecessary. The procedures for General Nodal Analysis can be summarized as follows:

1) Choose a node as the reference and assign every other node a voltage signal.
2) Form a tree including all voltage sources.

3) For each voltage source, write down the corresponding equation that relates the voltage source to the node voltages.
4) For each fundamental cut-set containing no voltage source, write down the corresponding cut-set equation.

Let's illustrate the procedures with the following example.

Example 4.3 Using the node voltages assigned for the circuit shown in **Figure 4.6**, write down the cut-set equations for the fundamental cut-sets associated with the chosen tree.

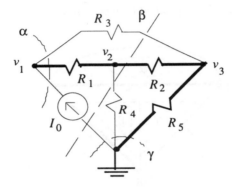

Figure 4.6. A circuit with a tree identified and the fundamental cut-sets

In the circuit, there are four nodes. Hence, a tree contains three branches. The tree containing R_1, R_2, and R_5 is identified in bold-face. Three fundamental cut-sets α, β, and γ are identified with dashed lines. The corresponding cut-set equations are

$$\frac{v_1 - v_2}{R_1} + \frac{v_1 - v_3}{R_3} = I_0$$

$$\frac{v_2 - v_3}{R_2} + \frac{v_1 - v_3}{R_3} + \frac{v_2}{R_4} = I_0$$

$$\frac{v_2}{R_4} + \frac{v_3}{R_5} = I_0$$

Example 4.4 Using the node voltages assigned for the circuit shown in **Figure 4.7**, write down the network equations which are necessary for solving the node voltages.

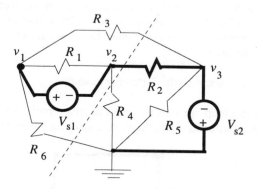

Figure 4.7. A circuit with a tree identified and a fundamental cut-set

In the circuit of **Figure 4.7**, there are four nodes. Hence, a tree contains three branches. The tree containing V_{s1}, V_{s2}, and R_2 is identified with bold-face. Of these branches, R_2 is the only one that is not a voltage source. Consequently, only the fundamental cut-set containing R_2 is shown. For the two voltage sources, we have

$$v_1 - v_2 = V_{s1}$$

$$v_3 = -V_{s2}$$

For the fundamental cut set shown, we have

$$\frac{v_1}{R_6} + \frac{v_1 - v_3}{R_3} + \frac{v_2}{R_4} + \frac{v_2 - v_3}{R_2} = 0$$

4.3 General Loop Analysis

With the help of network topology, the loop analysis can also be structured in such a way that is independent of whether current sources are present. The utilization of fundamental loops makes the forming of a superloop unnecessary. The procedure can be summarized as follows:

1) Form a tree excluding all current sources.

2) Identify each fundamental loop with a loop current.
3) For each fundamental loop containing a current source, write down the corresponding equation that relates the loop current to the assigned current signal of the source.
4) For each fundamental loop containing no current source, write down the corresponding loop equation via KVL. Note that the current in a branch is equal to the algebraic sum of the loop currents of those loops containing the branch.

Let's illustrate the procedure with the following example.

Example 4.5 Using the loop currents assigned for the circuit shown in **Figure 4.7**, write down the network equations which are necessary for solving the loop currents.

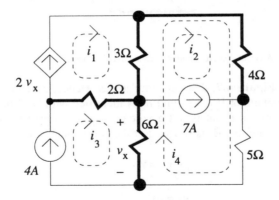

Figure 4.7. A circuit with a tree identified and the fundmental loops

There are five nodes and eight branches in the circuit. Hence, a tree contains four branches. There are four links. For each link, there is a corresponding fundamental loop. Fundamental loops are identified with dotted closed paths with directional signs and corresponding current signals. Three fundamental loops contain a current source each. For these, we have

$$i_1 = 2v_x = 2 \times 6(i_3 - i_4)$$

$$i_2 = -7 \text{ A}$$

$$i_3 = 4 \text{ A}$$

There is only one fundamental loop containing no current source, i.e., the loop identified by i_4. Applying KVL to this loop, we have

$$3(i_4 + i_2 - i_1) + 4(i_4 + i_2) + 5i_4 + 6(i_4 - i_3) = 0$$

Solving for i_4 and i_1, we find

$$i_4 = \frac{217}{54} \ A$$

$$i_1 = \frac{-2}{9} \ A$$

Problems

4.1-1 For the graph shown in **Figure 4.8**, determine the number of tree branches and the number of links.

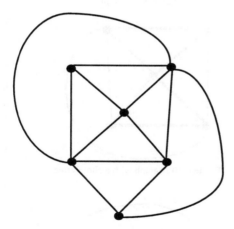

Figure 4.8 A graph

4.1-2 Determine the number of tree branches and the number of links for the graph shown in **Figure 4.9**. Identify a tree of the graph.

4.1-3 With the specified tree (consisting of heavy branches) for the graph shown in **Figure 4.10**, identify all the fundamental cut-sets.

4.1-4 With the specified tree (consisting of heavy branches) for the graph shown in **Figure 4.11**, identify all the fundamental cut-sets.

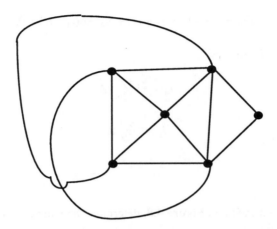

Figure 4.9 A graph

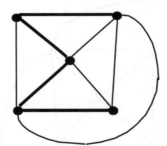

Figure 4.10 A graph with a chosen tree

Figure 4.11 A graph with a chosen tree

4.1-5 With the specified tree (consisting of heavy branches) for the graph shown in **Figure 4.12**, identify all the fundamental cut-sets.

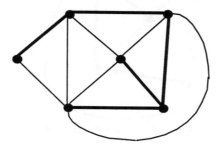

Figure 4.12 A graph with a chosen tree

4.1-6 With the specified tree (consisting of heavy branches) for the graph shown in **Figure 4.13**, identify all the fundamental loops.

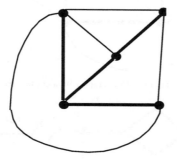

Figure 4.13 A graph with a chosen tree

4.1-7 With the specified tree (consisting of heavy branches) for the graph shown in **Figure 4.14**, identify all the fundamental loops.

4.1-8 With the specified tree (consisting of heavy branches) for the graph shown in **Figure 4.15**, identify all the fundamental loops.

4.2-1 Using the node voltages assigned and the general nodal analysis with the tree identified (consisting of heavy branches) for the circuit shown in **Figure 4.16**, write down the network equations necessary for solving these node voltages.

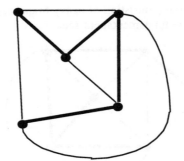

Figure 4.14 A graph with a chosen tree

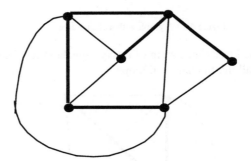

Figure 4.15 A graph with a chosen tree

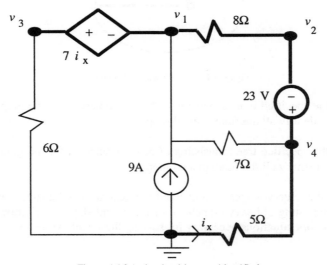

Figure 4.16 A circuit with a tree identified

4.2-2 Using the general nodal analysis with the tree identified for the circuit shown in **Figure 4.17**, find the current I_o.

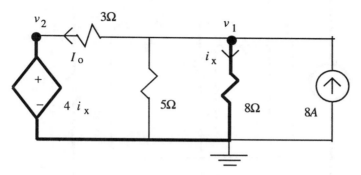

Figure 4.17 A circuit with a tree identified

4.3-1 Using the loop currents assigned and the general loop analysis with the tree identified for the circuit shown in **Figure 4.18**, write down the network equations necessary for solving these loop currents.

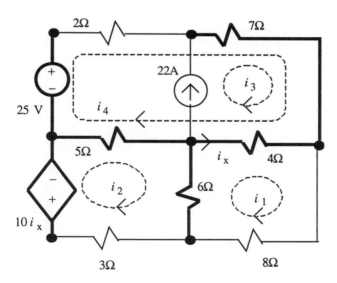

Figure 4.18 A circuit with a tree identified

4.3-2 Using the general loop analysis with the tree identified for the circuit shown in **Figure 4.19**, find the voltage V_o.

82

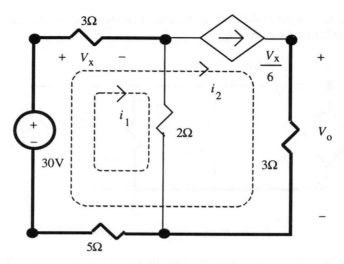

Figure 4.19 A circuit with a tree identified

Chapter 5
State Variable Method for RLC Circuits

In a given circuit, the knowledge of the voltage signals across capacitors and the current signals through inductors are sufficient enough for us to calculate other circuit quantities using only algebraic equations. Capacitor voltages and inductor currents are generally continuous in nature and are often chosen as the so-called state variables of the circuit. State variables are governed by a set of state equations that are first order differential equations. The coefficients in the state equations for a given circuit can be found via DC analysis of a SPICE program.

5.1 First Order Circuits

A circuit with only one energy storage element (an inductor or a capacitor but not both) is classified as a first-order circuit since this type of circuits is governed by a first order linear ordinary differential equation. A first-order circuit can further be classified as either an RC circuit (containing one capacitor) or an RL circuit (containing one inductor.)

5.1.1 RC Circuits

A circuit containing only one capacitor and no inductor is known as a first-order RC circuit. With the help of Thevenin's Theorem, the dynamic of such a circuit can be found by studying the circuit in **Figure 5.1**.

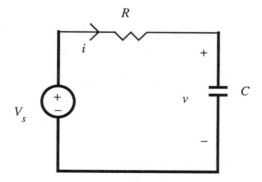

Figure 5.1. A first-order RC circuit

There are three nodes in the circuit. Hence, a tree consists of two branches (or elements.) Using the voltage source and the capacitor, a tree is formed and identified with heavy lines. Let us assign $v(t)$ as the voltage across the capacitor with polarity as shown while assign $i(t)$ as the current through the resistor with the given direction. Using the resistor, which is a link, to form a fundamental loop (the only loop in the circuit), we get the following loop equation.

$$v_s(t) = i(t)R + v(t) \qquad 5.1\text{-}1$$

Using the capacitor, which is a tree branch, to form a fundamental cut-set, we get the following cut-set equation

$$C\frac{dv}{dt} = i(t) \qquad 5.1\text{-}2$$

Using an expression for $i(t)$, obtained from **Equation 5.1-1**, we can turn **Equation 5.1-2** into the following

$$\frac{dv}{dt} = \frac{i}{C} = \frac{1}{C}\left(\frac{v_s - v}{R}\right) = \frac{-v}{CR} + \frac{v_s}{CR} \qquad 5.1\text{-}3$$

Equation 5.1-3 is the so-called state equation of the circuit with $v(t)$ chosen as the state variable.

The solution of **Equation 5.1-3** can be written as the summation of two parts: the homogeneous solution v_h and the particular solution v_p. The *homogeneous solution* satisfies the state equation with the source v_s set to zero, i.e.,

$$\frac{dv_h}{dt} = -\frac{v_h}{RC}$$

It is well known that we can assume that v_h takes the form of an exponential function as follows:

$$v_h = He^{st}$$

where H and s are constants to be determined.

Substituting the homogeneous solution v_h into the homogeneous equation, we obtain

$$sHe^{st} = -\frac{He^{st}}{RC}$$

For a nontrivial solution (i.e., $H \neq 0$), we must have

$$s = -\frac{1}{RC}$$

The product RC has of the same units as the time variable t. Hence, it is referred to as the time-constant of the circuit. After each period of a time-constant, the magnitude of v_h decreases by a factor of $1/e$. When a total period of several time-constants has past, the value of v_h can be approximated by zero. Consequently, this type of homogeneous solution is often referred to as the transient.

A *particular solution* is a specific solution of the original differential equation. A particular solution is often chosen to be the simplest one for a given source function v_s. There are three cases of interest:

1) v_s is a constant,
2) v_s is an exponential function,
3) v_s is a sinusoid.

Since a particular solution in either Case 1 or Case 3 does not approach to zero as time t goes to infinity, a particular solution is often referred to as the steady-state response.

Case 1. When v_s is a constant V_o, i.e., a DC source, we can assume that v_p is also a constant K. Substituting v_p in **Equation 5.1-3**, we have

$$0 = -\frac{K}{RC} + \frac{V_o}{RC}$$

Consequently,

$$v_p = K = V_o$$

Case 2. When v_s is an exponential function, say $V_o e^{bt}$, we can assume that v_p is of the same form, i.e., Ke^{bt}. Substituting this function in **Equation 5.1-3**, we have

$$bKe^{bt} = -\frac{Ke^{bt}}{RC} + \frac{V_o e^{bt}}{RC}$$

Solving for K, we have

$$K = \frac{V_o}{RCb+1}$$

Note that the particular solution is good only if b does not equal to $-1/(RC)$. If b equals to $-1/(RC)$, the particular solution is assumed to be of the following form

$$v_p = Kte^{bt}$$

where K is found to be $V_o/(RC)$.

Case 3. When v_s is a sinusoid, say $V_o \cos(\omega t + \delta)$, we assume that v_p is of the same form, i.e., $K \cos(\omega t + \gamma)$. Substituting this solution in **Equation 5.1-3**, we have

$$-\omega K \sin(\omega \cdot t + \gamma) = -\frac{K \cos(\omega \cdot t + \gamma)}{RC} + \frac{V_o \cos(\omega \cdot t + \delta)}{RC}$$

Solving for K and γ, we have

$$K = \frac{V_o}{\sqrt{1+(\omega RC)^2}}$$

$$\gamma = \delta - \tan^{-1}(\omega RC)$$

With v_h and v_p found, the complete solution for **Equation 5.1-3** is given by

$$v = v_h + v_p = He^{-\sigma t} + v_p$$

where $\sigma = 1/(RC)$. The constant H can then be determined with the initial value of v, i.e. $v(0^+)$ which is the limit of $v(t)$ as t approach to zero from the positive side of zero. We have

$$H = v(0^+) - v_p(0^+) = v(0^-) - v_p(0^+)$$

Being the capacitor voltage, $v(t)$ cannot be changed instantaneously. Hence, $v(0^+)$ equals to $v(0^-)$. The capacitor voltage $v(0^-)$ can be found by studying the circuit before $t=0$. If steady-state is assumed before $t=0$, the capacitor can be replaced with an open-circuit. With the capacitor replaced by an open-circuit, the RC circuit becomes a resistive circuit. An analysis of this resistive circuit leads to the value of the capacitor voltage. Let's examine the circuit in the following example.

Example 5.1.1. For $t < 0$, the capacitor voltage in **Figure 5.2** has been discharged to zero volts. Find the voltage $v(t)$ and the current $i(t)$ for $t > 0$.

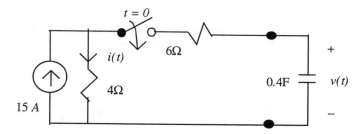

Figure 5.2 A first-order RC circuit

Solution: For $t \geq 0$ and for the voltage across the capacitor $v(t)$, the circuit in **Figure 5.2** is equivalent to the following circuit.

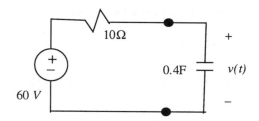

Figure 5.3 The equivalent circuit for finding $v(t)$ in **Figure 5.2**

From **Equation 5.1-3**, we have

$$\frac{dv}{dt} = -\frac{v}{4} + \frac{60}{4}$$

The solution is thus found to be

$$v(t) = v_h + v_p = He^{-t/4} + 60$$

where

$$H = v(0^+) - v_p(0^+) = v(0^-) - 60 = 0 - 60 = -60$$

Therefore, we have

$$v(t) = -60e^{-t/4} + 60 \ V$$

Once the voltage across the capacitor is found, we can replace the capacitor with a voltage source to find other quantities such as the current $i(t)$ as shown in **Figure 5.4**.

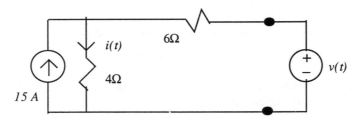

Figure 5.4 The equivalent circuit for other quantities in **Figure 5.2**

Applying superposition, we have

$$i(t) = \frac{6}{6+4} \times 15 + \frac{v(t)}{6+4} = 9 + (-6e^{-t/4} + 6) = 15 - 6e^{-t/4} \ A$$

Note that the current $i(t)$ is not continuous at time $t = 0$.

5.1.2 RL Circuits

A circuit containing only one inductor and no capacitor is known as a first-order RL circuit. With the help of Norton's Theorem, the characteristics of such a circuit can be found by studying circuit in **Figure 5.5**.

There are only two nodes in the circuit. Using only the resistor, a tree is formed. Let us assign $v(t)$ as the voltage across the resister with polarity as shown and assign $i(t)$ as the current through the inductor with the given direction. Using the resistor, which is a tree branch, to form a fundamental cut-set (the only one in the circuit), we get the following cut-set equation

$$i_s(t) = \frac{v(t)}{R} + i(t) \qquad 5.1\text{-}4$$

Using the inductor, which is a link, to form a fundamental loop, we get the following loop equation

$$L\frac{di}{dt} = v(t) \qquad 5.1\text{-}5$$

Using an expression for $v(t)$, obtained from **Equation 5.1-4**, we can turn **Equation 5.1-5** into

$$\frac{di}{dt} = \frac{v}{L} = \frac{R(i_s - i)}{L} = -\frac{Ri}{L} + \frac{Ri_s}{L} \qquad 5.1\text{-}6$$

Equation 5.1-6 is the so-called state equation of the circuit with $i(t)$ chosen as the state variable.

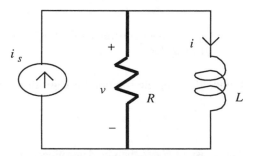

Figure 5.5 A first-order RL circuit

The solving of **Equation 5.1-6** is similar to that of **Equation 5.1-3**. Hence $i(t)$ can be written as the summation of two parts: the homogeneous solution i_h and the particular solution i_p. The homogeneous solution takes the following form

$$i_h = He^{st}$$

where

$$s = -R\!\!\Big/\!\!L$$

The time-constant of the circuit is L/R.

There are three cases of interests for the particular solution as well:

Case 1. If i_s is a constant, say I_o, then

$$i_s = I_o$$

Case 2. If i_s is an exponential function, say $I_o e^{bt}$, then

$$i_p = \frac{I_o}{1 + bL/R} e^{bt}$$

when b does not equal to $-R/L$. Should b equals to $-R/L$, we have

$$i_p = I_o \frac{R}{L} t e^{bt}$$

Case 3. If i_s is a sinusoid, say $I_o \cos(\omega t + \delta)$,

$$i_p = \frac{I_o}{\sqrt{1 + (\omega L/R)^2}} \cos[\omega \cdot t + \delta - \tan^{-1}(\omega L/R)]$$

With i_h and i_p found, the complete solution for **Equation 5.1-6** is given by

$$i = i_h + i_p = He^{-\sigma t} + i_p$$

where $\sigma = 1/(RC)$. The constant H can then be determined with the initial value of i, i.e. $i(0)$ and it is given by

$$H = i(0^+) - i_p(0^+) = i(0^-) - i_p(0^+)$$

where $i(0^-)$ denotes the limit of $i(t)$ as t approach to zero from the left side of zero. Being the inductor current, $i(t)$ cannot be changed instantaneously. Hence, $i(0^+)$ equals to $i(0^-)$. The inductor current $i(0^-)$ can be found by studying the circuit before $t=0$. If steady-state is assumed before $t=0$, the inductor can be replaced with a short-circuit. An analysis of the remaining resistive circuit would lead to the value of the inductor current. Let's examine the circuit in the following example.

Example 5.1.2. The switch in the circuit shown has been opened for a long time and is closed at $t = 0$. Find the current $i(t)$ and the voltage $v(t)$ for $t \geq 0$.

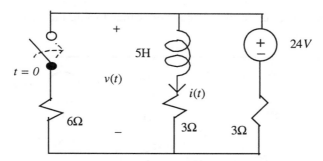

Figure 5.6 A first-order RL circuit

Solution: For $t \geq 0$ and for the current through the inductor $i(t)$, the circuit in **Figure 5.6** is equivalent to the following circuit through Norton's Theorem.

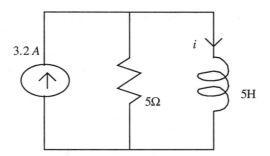

Figure 5.7 The equivalent circuit for $i(t)$ in **Figure 5.6**

From **Equation 5.1-6**, we have

$$\frac{di}{dt} = -i + 3.2$$

The solution is thus found to be

$$i(t) = i_h + i_p = He^{-t} + 3.2$$

where

$$H = i(0^+) - i_p(0^+) = i(0^-) - 3.2 = \frac{24}{3+3} - 3.2 = 0.8\,A$$

Note that the initial current in the inductor is found using the equivalent circuit in **Figure 5.8**.

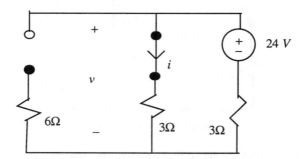

Figure 5.8 The equivalent circuit before $t = 0$ in **Figure 5.6**

Once the current in the inductor is found, we can replace the inductor with a current source to find other quantities such as the voltage $v(t)$ as shown in **Figure 5.9**.

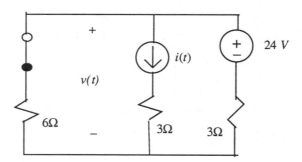

Figure 5.9 The equivalent circuit for other quantities in **Figure 5.6**

Applying superposition, we have

$$v(t) = \frac{6}{3+6} \times 24 - \frac{6 \times 3}{6+3} \times i(t) = 16 - 2 \times (0.8e^{-t} + 3.2) = 9.6 - 1.6e^{-t} \; V$$

Note that the voltage $v(t)$ is not continuous at time $t = 0$.

5.2 Higher Order Circuits

When a RLC circuit contains two or more energy-storage elements, the topological arguments in **Section 5.1** can be applied directly to get the state equations of the circuit. However, the inclusion of Thevenin's and Norton's Theorems in the analysis offers an efficient alternative as well as desirable outcomes.

5.2.1 Topological Method

The procedures are outlined as follows:

(i) Form a tree, including all the voltage sources and all the capacitors but excluding all current sources and all inductors[*].

(ii) For each passive element in the tree, assign a voltage signal.

(iii) For each passive element in the co-tree, assign a current signal.

(iv) Using each passive element (a resistor or a capacitor) in the tree, we form a fundamental cut-set and write down the corresponding cut-set equation. There are two distinctive possibilities. For a resistor in the tree (referred as a tree resistor from now on), the current through it is given by

$$\frac{v_{Rl}}{R_l} = \sum (-1)^{\alpha_{li}} i_{si} + \sum (-1)^{\beta_{lj}} i_{Rj} + \sum (-1)^{\gamma_{lk}} i_{Lk} \qquad 5.2\text{-}1$$

In the equation, the summation is over those circuit elements in the fundamental cut-set excluding the tree resistor. The subscript s denotes that the current is from a current source, the subscript L denotes that

---

[*] Note that if such a tree can not be formed, the circuit contains capacitors, inductors, voltage sources or current sources that are not independent.

the current is from an inductor, and the subscript R denotes that the current is from a resistor in the co-tree. The power of -1, i.e., α_{li}, β_{lj}, or γ_{lj}, is either odd or even depending on whether the corresponding current is in the same direction of the current of the tree resistor or not. For a capacitor in the tree, the current through it is given by

$$C_m \frac{dv_{Cm}}{dt} = \sum (-1)^{\delta_{mi}} i_{si} + \sum (-1)^{\sigma_{mj}} i_{Rj} + \sum (-1)^{\omega_{mk}} i_{Lk} \qquad 5.2\text{-}2$$

(v) Using each passive element (an inductor or a resistor) in the co-tree, we form a fundamental loop and write down the corresponding loop equation. There are also two distinctive possibilities. For a resistor in the co-tree (referred as a link resistor from now on), the voltage across it is given by

$$i_{Rj} Rj = \sum (-1)^{\rho_{jn}} v_{sn} + \sum (-1)^{\phi_{jl}} v_{Rl} + \sum (-1)^{\psi_{jm}} v_{Cm} \qquad 5.2\text{-}3$$

In the equation, the summation is over those circuit elements in the fundamental loop excluding the link resistor. The subscript s denotes that the voltage is from a voltage source while the subscript R denotes that the voltage is from a resistor in the tree. The power of -1, i.e. ρ_{li}, ϕ_{li} or ψ_{lj}, is either odd or even depending on whether the corresponding voltage is of the same polarity as the voltage of the link resistor or not.

For an inductor in the tree, the voltage across it is given by

$$L_k \frac{di_{Lk}}{dt} = \sum (-1)^{\eta_{kn}} v_{sn} + \sum (-1)^{\mu_{kl}} v_{Rl} + \sum (-1)^{\xi_{km}} v_{Cm} \qquad 5.2\text{-}4$$

(vi) For each resistor in the circuit, there is a corresponding equation such as **Equation 5.2-1** or **Equation 5.2-3**. From these equations, tree resistor voltages and link resistor currents can be solved as linear combinations of source voltages, source currents, inductor currents, and capacitor voltages. By replacing tree resistor voltages and link resistor currents with these linear combinations in **Equations 5.2-2** and **5.2-4**, state equations of the circuit are thus found.

To illustrate the procedure, let us examine the circuit shown in **Figure 5.10**. The chosen tree, identified with heavy lines, consists of the voltage source V_s, the capacitor C, and the resistor R_2. Passive elements in the tree are assigned with voltage signals v_C and v_2 while passive elements in the co-tree are assigned with current signals i_1 and i_L.

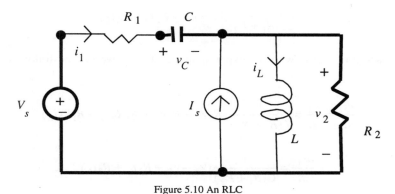

Figure 5.10 An RLC

Using the resistor R_2 to form a fundamental cut-set, we have the following cut-set equation

$$\frac{v_2}{R_2} = I_s + i_1 - i_L$$

Using the capacitor C to form another fundamental cut-set, we have

$$C\frac{dv_C}{dt} = i_1$$

Using the resistor R_1 to form a fundamental loop, we have the following loop equation

$$i_1 R_1 = V_s - v_C - v_2$$

Using the inductor L to form another fundamental loop, we have

$$L\frac{di_L}{dt} = v_2$$

From the two algebraic equations, we get

$$v_2 = \frac{1}{R_1 + R_2}(R_2 V_s - R_2 v_C + R_1 R_2 I_s - R_1 R_2 i_L)$$

and

$$i_1 = \frac{1}{R_1 + R_2}(V_s - v_C - R_2 I_s + R_2 i_L)$$

By replacing v_2 and i_1 with the expressions just found, we get the following state equations

$$\frac{di_L}{dt} = \frac{v_2}{L} = \frac{1}{L(R_1 + R_2)}(R_2 V_s - R_2 v_C + R_1 R_2 I_s - R_1 R_2 i_L)$$

and

$$\frac{dv_C}{dt} = \frac{1}{C(R_1 + R_2)}(V_s - v_C - R_2 I_s + R_2 i_L)$$

Once the state equations are found, a standardized method can be applied to find the solution.

5.2.2 Generalized Thevenin's Approach

Based on the state-variable method, a generalized Thevenin's approach is proposed as follows:

First, assign each capacitor a voltage signal, say v_{Cj}, while assigning each inductor a current signal, say i_{Lj}. An RLC circuit with n energy storage elements is equivalent to n topologically disjoint first-order circuits as shown in **Figure 5.11**.

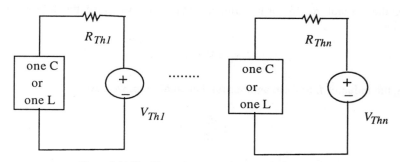

Figure 5.11 The Thevenin's equivalent circuit of an RLC

There are n pairs of Thevenin's equivalent voltages V_{Thj} and Thevenin's equivalent resistances R_{Thj}, one pair for each energy-storage element. Each pair of (V_{Thj}, R_{Thj}) is found by replacing all the rest of energy-storage element with sources, inductors with current sources and capacitors with voltage sources. These converted sources are identified by the same signals assigned at the very beginning. Hence, the voltage V_{Thj} is a linear combination of the original voltage sources, the original current sources, and the converted sources excluding i_{Lj} or v_{Ci}. Once these Thevenin's equivalent quantities are found, the state equation for each first-order circuit can be written as shown in **Section 5.1**.

We have

$$\frac{dv_{Cj}}{dt} = -\frac{v_{Cj}}{C_j R_{Thj}} + \frac{V_{Thj}}{C_j R_{Thj}}$$

or

$$\frac{di_{Lj}}{dt} = -\frac{R_{Thj} i_{Lj}}{L_j} + \frac{V_{Thj}}{L_j}$$

Let us illustrate the procedure by studying the circuit in **Figure 5.12**. The equivalent circuit has two disjoint parts since there are only two energy-storage elements. Hence, only two pairs of Thevenin's equivalent quantities are needed.

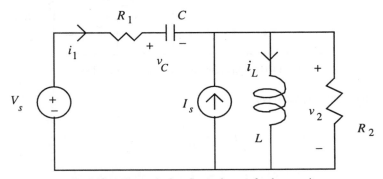

Figure 5.12 The equivalent first-order part for the capacitor

For V_{Th1} and R_{Th1}, we replace the inductor with a current source i_L. The circuit in **Figure 5.12** becomes the one in **Figure 5.13**.

Through analysis, we have

$$R_{Th1} = R_1 + R_2, \quad V_{Th1} = V_s - R_2(I_s - i_L)$$

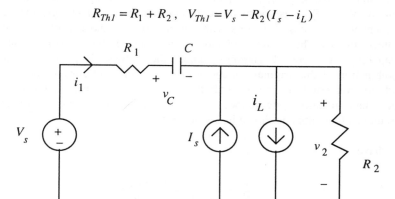

Figure 5.13 The equivalent first-order part for the capacitor

For V_{Th2} and R_{Th2}, we replace the capacitor with a voltage source v_C. The circuit becomes

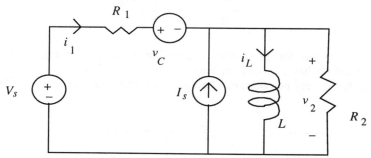

Figure 5.14 The equivalent first-order part for the inductor

Through analysis, we have

$$R_{Th2} = \frac{R_1 R_2}{R_1 + R_2}$$

$$V_{Th2} = V_s \frac{R_2}{R_1 + R_2} - v_C \frac{R_2}{R_1 + R_2} + I_s \frac{R_1 R_2}{R_1 + R_2}$$

Furthermore, we have

$$\frac{dv_C}{dt} = -\frac{v_C}{CR_{Th1}} + \frac{V_{Th1}}{CR_{Th1}} = -\frac{v_C}{C(R_1 + R_2)} + \frac{V_s - R_2(I_s - i_L)}{C(R_1 + R_2)}$$

and

$$\frac{di_L}{dt} = -\frac{R_{Th2}i_L}{L} + \frac{V_{Th2}}{L} = -\frac{R_1 R_2 i_L}{L(R_1 + R_2)} + \frac{R_2 V_s - R_2 v_C + R_1 R_2 I_s}{L(R_1 + R_2)}$$

These two equations are the state equations derived in **Section 5.2.1**.

5.2.3 Generalized Norton's Approach

As in **Section 5.2.2**, a generalized Norton's approach is proposed as follows:

First, assign each capacitor a voltage signal, say v_{Cj}, while assigning each inductor a current signal, say i_{Lj}. An RLC circuit with n energy storage elements is equivalent to n topological disjoint first-order circuits as shown in **Figure 5.15**.

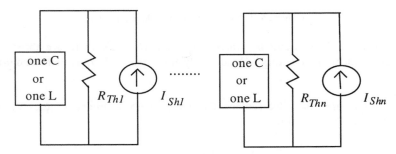

Figure 5.15 The Norton's equivalent circuit of an RLC

There are n pairs of short-circuit currents I_{Shj} and Thevenin's equivalent resistances R_{Thj}, one pair for each energy-storage element. Each pair of (I_{Shj}, R_{Thj}) is found by replacing all the rest of energy-storage element with sources, inductors with current sources and capacitors with voltage sources. These converted sources are identified by the same signal assigned at the very beginning. Hence, the current I_{Shj} is a linear combination of the original voltage sources, the original current sources, and the converted sources excluding i_{Lj} or v_{Ci}. Once these equivalent quantities are found, the state equation for each first-order circuit can be written as shown in **Section 5.1**. We have

$$\frac{dv_{Cj}}{dt} = -\frac{v_{Cj}}{C_j R_{Thj}} + \frac{I_{Shj}}{C_j}$$

or

$$\frac{di_{Lj}}{dt} = -\frac{R_{Thj} i_{Lj}}{L_j} + \frac{R_{Thj} I_{Shj}}{L_j}$$

Let us illustrate the procedure by studying the circuit in **Figure 5.16**.

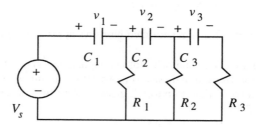

Figure 5.16 An RLC circuit with three capacitors

The equivalent circuit has three disjoint parts since there are three energy-storage elements (capacitors.) Hence, only three pairs of equivalent quantities are needed.

For I_{Sh1} and R_{Th1}, we replace C_2 and C_3 with a voltage sources v_2 and v_3, respectively. The circuit becomes

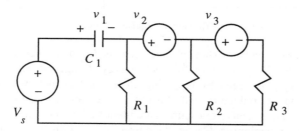

Figure 5.17 The equivalent first-order part for the first capacitor

Through analysis, we have

$$I_{Sh1} = V_s \left(\frac{1}{R_1} + \frac{1}{R_2} + \frac{1}{R_3}\right) - v_2 \left(\frac{1}{R_2} + \frac{1}{R_3}\right) - v_3 \frac{1}{R_3}$$

and

$$R_{Th1} = (\frac{1}{R_1} + \frac{1}{R_2} + \frac{1}{R_3})^{-1}$$

For I_{Sh2} and R_{Th2}, we replace C_1 and C_3 with a voltage sources v_1 and v_3, respectively. The circuit becomes

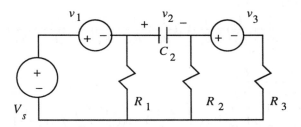

Figure 5.18 The equivalent first-order part for the second capacitor

Through analysis, we have

$$I_{Sh2} = V_s (\frac{1}{R_2} + \frac{1}{R_3}) - v_1 (\frac{1}{R_2} + \frac{1}{R_3}) - v_3 \frac{1}{R_3}$$

$$R_{Th2} = (\frac{1}{R_2} + \frac{1}{R_3})^{-1}$$

For I_{sh3} and R_{Th3}, we replace C_1 and C_2 with a voltage sources v_1 and v_2, respectively. The circuit becomes

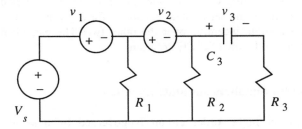

Figure 5.19 The equivalent first-order part for the third capacitor

Through analysis, we have

$$I_{Sh3} = V_s \frac{1}{R_3} - v_1 \frac{1}{R_3} - v_2 \frac{1}{R_3} \qquad \text{and} \qquad R_{Th3} = R_3$$

Furthermore, we have

$$\frac{dv_1}{dt} = -\frac{v_1}{C_1 R_{Th1}} + \frac{I_{Sh1}}{C_1}$$

$$= -\frac{v_1}{C_1}(\frac{1}{R_1} + \frac{1}{R_2} + \frac{1}{R_3}) + \frac{1}{C_1}[V_s(\frac{1}{R_1} + \frac{1}{R_2} + \frac{1}{R_3}) - v_2(\frac{1}{R_2} + \frac{1}{R_3}) - v_3 \frac{1}{R_3}]$$

$$\frac{dv_2}{dt} = -\frac{v_2}{C_2 R_{Th2}} + \frac{I_{Sh2}}{C_2}$$

$$= -\frac{v_2}{C_2}(\frac{1}{R_2} + \frac{1}{R_3}) + \frac{1}{C_2}[V_s(\frac{1}{R_2} + \frac{1}{R_3}) - v_1(\frac{1}{R_2} + \frac{1}{R_3}) - v_3 \frac{1}{R_3}]$$

$$\frac{dv_3}{dt} = -\frac{v_3}{C_3 R_{Th3}} + \frac{I_{Sh3}}{C_3} = -\frac{v_3}{C_3 R_3} + \frac{1}{C_3}[V_s \frac{1}{R_3} - v_1 \frac{1}{R_3} - v_2 \frac{1}{R_3}]$$

5.2.4 Canonical Approach

An alternative to either the generalized Thevenin's approach or the generalized Norton's approach is made possible when Thevenin's equivalent part is reserved for a capacitor while Norton's equivalent part is reserved for an inductor. Consequently, the third equivalent circuit for an RLC is formed. The equivalent circuit is as shown in **Figure 5.20**.

Except possible differences in the ordering of state-variables, all three approaches yield the same set of state equations for a given circuit. The canonical approach offers a slight incentive over the other two. The canonical approach can easily be implemented with a SPICE Program.

5.3 Solution of the Simultaneous State Equations

A second- or higher-order circuit is governed by a set of two or more simultaneous state equations. The solution for such a set of equations can be carried out in a similar fashion as in **Section 5.1**. However, modifications are necessary since there are two or more simultaneous equations.

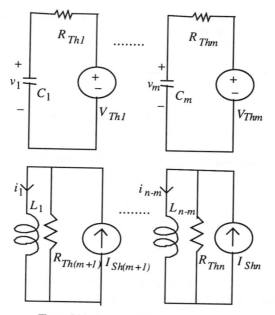

Figure 5.20 The canonical equivalent circuit

5.3.1 Second-order Circuits

The state equations for a second-order circuit consists of two simultaneous first order linear differential equations as demonstrated with the circuit in **Figure 5.3**. These equations can be generalized as follows:

$$
\begin{cases}
\dfrac{dx_1}{dt} = a_{11}x_1 + a_{12}x_2 + f_1(t) \\[3mm]
\dfrac{dx_2}{dt} = a_{21}x_1 + a_{22}x_2 + f_2(t)
\end{cases}
$$

$$5.3\text{-}1$$

In **Equation 5.3-1**, variables x_1 and x_2 are referred to as state variables representing capacitor voltages or inductor currents. The forcing functions f_1 and f_2 are due to the independent sources in the circuit.

5.3.1.1 Homogeneous Solutions

The solution of each state variable in **Equation 5.3-1** can be written as the summation of two parts: the homogeneous solution x_{kh} and the particular solution

x_{kp}, with $k = 1$ or 2. The *homogeneous solution parts* satisfy the state equations with the forcing functions set to zero, i.e.,

$$\begin{cases} \dfrac{dx_{1h}}{dt} = a_{11}x_{1h} + a_{12}x_{2h} \\ \\ \dfrac{dx_{2h}}{dt} = a_{21}x_{1h} + a_{22}x_{2h} \end{cases}$$

5.3-2

The homogenous solutions are assumed to take the form of exponential functions as follows:

$$\begin{cases} x_{1h} = H_1 e^{st} \\ \\ x_{2h} = H_2 e^{st} \end{cases}$$

where H_1, H_2 and s are constants to be determined. With the homogeneous solutions shown, **Equations 5.3.2** become the following

$$\begin{cases} sH_1 e^{st} = a_{11}H_1 e^{st} + a_{12}H_2 e^{st} \\ \\ sH_2 e^{st} = a_{21}H_1 e^{st} + a_{22}H_2 e^{st} \end{cases}$$

5.3-3

With matrix notation, **Equations 5.3.3** become the following

$$\begin{bmatrix} s-a_{11} & -a_{12} \\ -a_{21} & s-a_{22} \end{bmatrix} \begin{bmatrix} H_1 \\ H_2 \end{bmatrix} e^{st} = 0$$

5.3-4

For a non-trivial solution, the determinant of the 2 by 2 matrix in **Equation 5.3-4** must be zero. Namely, we must have

$$\begin{vmatrix} s-a_{11} & -a_{12} \\ -a_{21} & s-a_{22} \end{vmatrix} = (s-a_{11})(s-a_{22}) - a_{12}a_{21} = 0$$

5.3-5

Equation 5.3-5 is known as the characteristic equation of the second-order circuit. This equation can be written as follows:

$$s^2 + 2\xi\omega_n s + \omega_n^2 = 0 \qquad\qquad \text{5.3-6}$$

with

$$\begin{cases} \omega_n^2 = a_{11}a_{22} - a_{12}a_{21} \\[2mm] 2\xi\omega_n = -a_{11} - a_{22} \end{cases}$$

Equation 5.3-6 has two solutions. They are

$$\begin{cases} s_1 = -\left(\xi + \sqrt{\xi^2 - 1}\right) \cdot \omega_n \\[3mm] s_2 = -\left(\xi - \sqrt{\xi^2 - 1}\right) \cdot \omega_n \end{cases} \qquad\qquad \text{5.3-7}$$

The constant ξ is normally non-negative and is known as the damping ratio. The value of the damping ratio effects the characteristics of the circuit. There are four cases of interest.

Case 1: When $\xi = 0$, we have two imaginary roots for the characteristic equation as follows:

$$s_1 = -s_2 = -j\omega_n$$

The homogeneous solutions of the state variable take the following form

$$\begin{aligned} x_{kh} &= H_{k1}e^{-j\omega_n t} + H_{k2}e^{j\omega_n t} \\ &= A_{k1}\cos\omega_n t + A_{k2}\sin\omega_n t = \rho_k \cos(\omega_n t + \phi_k) \end{aligned} \qquad \text{5.3-8}$$

with $k = 1$ or 2. **Equation 5.3-7** represents a sinusoidal function with a fixed amplitude. This is known as an undamped response. The corresponding angular frequency ω_n is known as undamped natural frequency.

Case 2: When $1 > \xi > 0$, we have

$$s_1 = -\sigma - j\omega \quad \text{and} \quad s_2 = -\sigma + j\omega$$

with

$$\sigma = \xi\omega_n \quad \text{and} \quad \omega = \omega_n\sqrt{1-\xi^2}$$

The homogeneous solutions of the state variables take the following form

$$x_{kh} = H_{k1}e^{-\sigma-j\omega t} + H_{k2}e^{-\sigma+j\omega t}$$
$$= e^{-\sigma t}\left(A_{k1}\cos\omega t + A_{k2}\sin\omega t\right) = \rho_k e^{-\sigma t}\cos(\omega t + \phi_k)$$

5.3-9

with $k = 1$ or 2. **Equation 5.3-9** can be considered as a sinusoidal function with a decreasing amplitude. This is known as an under-damped response.

Case 3: When $\xi = 1$, we have

$$s_1 = s_2 = -\sigma$$

The homogeneous solutions of the state variable take the following form

$$x_{kh} = H_{k1}e^{-\sigma t} + H_{k2}te^{-\sigma t}$$

5.3-10

with $k = 1$ or 2. This is known as a critically-damped response.

Case 4: When $\xi > 1$, the roots of the characteristic equation are two unequal negative real number. The homogeneous solutions of the state variable take the following form

$$x_{kh} = H_{k1}e^{s_1 t} + H_{k2}e^{s_2 t}$$

5.3-11

with $k = 1$ or 2. This is known as an over-damped response.

Except in *Case 1*, the homogeneous solutions approach to zero as time t goes to infinity. Hence, the homogeneous solutions are referred to as the transient. The coefficients in the homogeneous solutions in each case listed above are not independent. They are governed by **Equation 5.3-4** as demonstrated in **Example 5.3.1**.

Example 5.3.1 In **Figure 5.21**, a resistor, an inductor and a capacitor are connected in series in a loop. At $t = 0$, the voltage $v(t)$ across the capacitor is 12V and the current $i(t)$ in the inductor is 9A. Find $v(t)$ and $i(t)$ for $t > 0$.

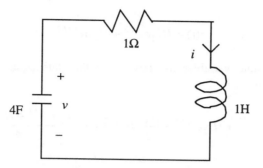

Figure 5.21 A second-order series RLC circuit

Solution: Using $v(t)$ and $i(t)$ as state variables, the state equations are found as follows:

$$\begin{cases} \dfrac{dv}{dt} = -\dfrac{i}{C} = -\dfrac{i}{4} \\[2mm] \dfrac{di}{dt} = \dfrac{v-iR}{L} = v-i \end{cases}$$

Note that the state equations contain no forcing functions. The solutions of the state equations consist of only homogeneous solutions. The characteristic equation is found to be as follows:

$$\begin{vmatrix} s & \dfrac{1}{4} \\[2mm] -1 & s+1 \end{vmatrix} = s(s+1)+\dfrac{1}{4} = (s+\dfrac{1}{2})^2 = 0$$

Hence, there are two identical roots for the characteristic equation. These roots are

$$s_1 = s_2 = -\dfrac{1}{2}$$

Therefore, the state variables are of the following form:

$$v(t) = H_{11}e^{-t/2} + H_{12}te^{-t/2}$$

and

$$i(t) = H_{21}e^{-t/2} + H_{22}te^{-t/2}$$

Substituting the state variables as shown into the state equations, we have the following equations:

$$-\frac{H_{11}}{2}e^{-t/2} + H_{12}e^{-t/2} - \frac{H_{12}}{2}te^{-t/2} = -\frac{H_{21}e^{-t/2} + H_{22}te^{-t/2}}{4}$$

and

$$-\frac{H_{21}}{2}e^{-t/2} + H_{22}e^{-t/2} - \frac{H_{22}}{2}te^{-t/2}$$
$$= H_{11}e^{-t/2} + H_{12}te^{-t/2} - (H_{21}e^{-t/2} + H_{22}te^{-t/2})$$

Collecting terms with common factors, we have

$$\left(-\frac{H_{11}}{2} + H_{12} + \frac{H_{21}}{4}\right) \cdot e^{-t/2} + \left(-\frac{H_{12}}{2} + \frac{H_{22}}{4}\right) \cdot te^{-t/2} = 0$$

and

$$\left(-\frac{H_{21}}{2} + H_{22} - H_{11} + H_{21}\right) \cdot e^{-t/2} + \left(-\frac{H_{22}}{2} - H_{12} + H_{22}\right) \cdot te^{-t/2} = 0$$

Since the two time functions are linearly independent, the coefficients of the time functions must be zero. Hence, we have

$$-\frac{H_{11}}{2} + H_{12} + \frac{H_{21}}{4} = 0$$

$$-\frac{H_{12}}{2} + \frac{H_{22}}{4} = 0$$

$$-\frac{H_{21}}{2} + H_{22} - H_{11} + H_{21} = 0$$

$$-\frac{H_{22}}{2} - H_{12} + H_{22} = 0$$

Although there are four equations, only two of them are linearly independent. From either the second or the fourth equation, we found that

$$H_{22} = 2H_{12}$$

From the first equation, we have

$$H_{12} = \frac{H_{11}}{2} - \frac{H_{21}}{4}$$

The third equation is redundant. Incorporating the initial conditions, we have

$$\begin{cases} v(0) = H_{11} = 12 \\ \\ i(0) = H_{21} = 9 \end{cases}$$

Consequently, we have

$$H_{12} = \frac{H_{11}}{2} - \frac{H_{21}}{4} = \frac{12}{2} - \frac{9}{4} = \frac{15}{4}$$

$$H_{22} = 2H_{12} = 2 \times \frac{15}{4} = \frac{15}{2}$$

The solutions for the state variables are thus as follows:

$$\begin{cases} v(t) = 12e^{-t/2} + \frac{15}{4}te^{-t/2} \\ \\ i(t) = 9e^{-t/2} + \frac{15}{2}te^{-t/2} \end{cases}$$

for $t > 0$.

5.3.1.2 Particular Solutions

A *particular solution* of a state variable is part of a set of specific solutions of the original state equations, **Equations 5.3-1**. Similar to the procedures outlined in **Section 5.1**, a particular solution is often chosen to be the simplest one for a given source function as described in Section 5.1. The procedure is illustrated with the following example.

Example 5.3.2 Assume the circuit in **Figure 5.22** has reached steady state for t < 0. Find the voltage across the capacitor and the current through the inductor.

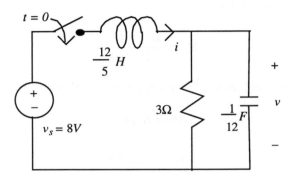

Figure 5.22 A second-order RLC circuit

Solution: Using $v(t)$ and $i(t)$ as state variables, the state equations are found as follows:

$$\begin{cases} \dfrac{dv}{dt} = \dfrac{1}{C}\left(i - \dfrac{v}{R}\right) = 12i - 4v \\[3mm] \dfrac{di}{dt} = \dfrac{v_s - v}{L} = \dfrac{10}{3} - \dfrac{5}{12}v \end{cases}$$

for $t > 0$.

Note that the state equations contain a constant forcing function. Set this forcing function to zero, we have

$$\begin{cases} \dfrac{dv_h}{dt} = 12i_h - 4v_h \\[3mm] \dfrac{di_h}{dt} = -\dfrac{5}{12}v_h \end{cases}$$

5.3-12

The characteristic equation is found to be

$$\begin{vmatrix} s+4 & -12 \\ \dfrac{5}{12} & s \end{vmatrix} = (s+4)s+5 = (s+2)^2 +1 = 0$$

Hence, the roots of the characteristic equation are a pair of complex conjugate numbers. These roots are

$$s_1 = -2 - j \quad \text{and} \quad s_2 = -2 + j$$

Therefore, the homogeneous solutions are of the following form

$$\begin{cases} v_h(t) = e^{-2t} \left(A_{11} \cos t + A_{12} \sin t \right) \\ \\ i_h(t) = e^{-2t} \left(A_{21} \cos t + A_{22} \sin t \right) \end{cases}$$

Substituting the homogeneous solutions as shown into **Equation 5.3-12**, we have the following equations:

$$-2e^{-2t} \left(A_{11} \cos t + A_{12} \sin t \right) + e^{-2t} \left(-A_{11} \sin t + A_{12} \cos t \right)$$
$$= -4e^{-2t} \left(A_{11} \cos t + A_{12} \sin t \right) + 12e^{-2t} \left(A_{21} \cos t + A_{22} \sin t \right)$$

and

$$-2e^{-2t} \left(A_{21} \cos t + A_{22} \sin t \right) + e^{-2t} \left(-A_{21} \sin t + A_{22} \cos t \right)$$
$$= \frac{-5}{12} e^{-2t} \left(A_{11} \cos t + A_{12} \sin t \right)$$

Collecting terms with common factors, we have

$$\left(-2A_{11} + A_{12} + 4A_{11} - 12A_{21} \right) \cdot e^{-2t} \cos t$$
$$+ \left(-2A_{12} - A_{11} + 4A_{12} - 12A_{22} \right) \cdot e^{-2t} \sin t = 0$$

and

$$\left(-2A_{21} + A_{22} + \frac{5}{12} A_{11} \right) \cdot e^{-2t} \cos t + \left(-2A_{22} - A_{21} + \frac{5}{12} A_{12} \right) \cdot e^{-2t} \sin t = 0$$

Since the two time-functions in each equation shown are linearly independent, the coefficients of the time functions must be zero. Hence, we have

$$-2A_{11} + A_{12} + 4A_{11} - 12A_{21} = 0$$

$$-2A_{12} - A_{11} + 4A_{12} - 12A_{22} = 0$$

$$-2A_{21} + A_{22} + \frac{5}{12}A_{11} = 0$$

$$-2A_{22} - A_{21} + \frac{5}{12}A_{12} = 0$$

Although there are four equations, only two of them are linearly independent. From the first equation, we found that

$$A_{12} = +2A_{11} - 4A_{11} + 12A_{21} = -2A_{11} + 12A_{21}$$

From the third equation, we found that

$$A_{22} = 2A_{21} - \frac{5}{12}A_{11}$$

Since there is a constant forcing function in the state equations, the particular solutions are assumed as follows:

$$v_p(t) = B_1 \quad \text{and} \quad i_p(t) = B_2$$

Substitute the particular solutions into the state equations, we have

$$\begin{cases} 0 = 12B_2 - 4B_1 \\ 0 = \dfrac{10}{3} - \dfrac{5}{12}B_1 \end{cases}$$

Solving for the constants, we have

$$B_1 = 8 \quad \text{and} \quad B_2 = \frac{B_1}{3} = \frac{8}{3}$$

The complete solutions for the state equations are as follows:

$$\begin{cases} v = v_h(t) + v_p(t) = e^{-2t}(A_{11}\cos t + A_{12}\sin t) + B_1 \\[2mm] i = i_h(t) + i_p(t) = e^{-2t}(A_{21}\cos t + A_{22}\sin t) + B_2 \end{cases}$$

When $t < 0$, the capacitor voltage and the inductor current are zero since there is no source connected to the circuit. At $t = 0$, we have

$$\begin{cases} v(0) = A_{11} + B_1 = 0 \\[2mm] i(0) = A_{21} + B_2 = 0 \end{cases}$$

Consequently, we have

$$A_{11} = -B_1 = -8 \quad \text{and} \quad A_{21} = -B_2 = \frac{-8}{3}$$

$$A_{12} = -2A_{11} + 12A_{21} = -2(-8) + 12\left(\frac{-8}{3}\right) = -16$$

$$A_{22} = 2A_{21} - \frac{5}{12}A_{11} = 2\left(\frac{-8}{3}\right) - \frac{5}{12}(-8) = -2$$

The solutions for the state variables are thus as follows:

$$\begin{cases} v = e^{-2t}(-8\cos t - 16\sin t) + 8 \\[2mm] i = e^{-2t}\left(\frac{-8}{3}\cos t - 2\sin t\right) + \frac{8}{3} \end{cases}$$

for $t > 0$.

5.3.2 Higher-order Circuits

Although a similar approach as in **Section 5.3.1** can be used to study third- or higher-order circuits, it becomes more difficult to keep track of the coefficients of the homogeneous solutions when there are three or more state variables. An approach based on Matrix Theory and the Laplace Transform enables us to simplify the notation and derivation. However, we feel that it is beyond the scope of this book. Readers are referred to the many textbooks in Signals & Systems or Automatic Control Theory for details.

Problems

5.1-1. The switch in the circuit shown in **Figure 5.23** has been opened for a long time and is closed at $t = 0$. Find the voltage $v(t)$ for $t \geq 0$.

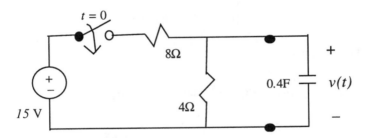

Figure 5.23 A first-order RC circuit

5.1-2. The switch of the circuit shown in **Figure 5.24** has been closed for a long time and is opened at $t = 0$. Find the voltage $v(t)$ for $t > 0$.

5.1-3. The switch in the circuit shown in **Figure 5.25** has been closed for a long time and is opened at $t = 0$. Find the voltage $v(t)$ for $t > 0$.

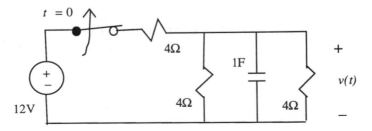

Figure 5.24 A first-order RC circuit

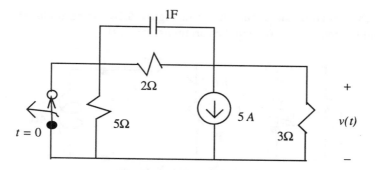

Figure 5.25 A first-order RC circuit

5.1-4. The switch of the circuit shown in **Figure 5.26** has been in position *a* for a long time. At $t = 0$ the switch moves instantaneously to position *b*. Find the current $i(t)$ for $t > 0$.

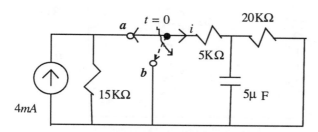

Figure 5.26 A first-order RC circuit

5.1-5. The switch of the circuit shown in **Figure 5.27** has been opened for a long time and is closed at $t = 0$. Find the voltage $v(t)$ for $t > 0$.

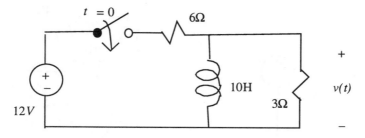

Figure 5.27 A first-order RL circuit

5.1-6. The switch of the circuit shown in **Figure 5.28** has been closed for a long time and is opened at $t = 0$. Find the current $i(t)$ for $t \geq 0$.

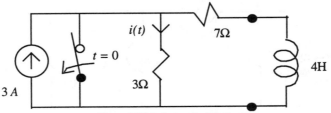

Figure 5.28 A first-order RL circuit

5.1-7. The switch of the circuit shown in **Figure 5.29** has been opened for a long time and is closed at $t = 0$. Find the current $i(t)$ for $t > 0$.

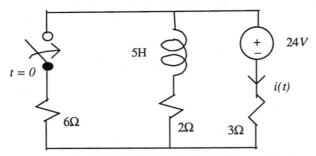

Figure 5.29 A first-order RL circuit

5.2-1. Using the quantities identified for the circuit shown in **Figure 5.30** as the state variables, find the state equations of the circuit.

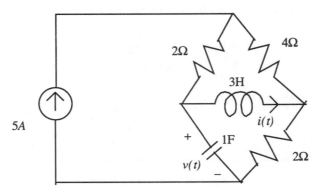

Figure 5.30 A second-order RLC circuit

5.2-2. Using the capacitor voltages and the inductor currents for the circuit shown in **Figure 5.31** as state variables, find the state equations of the circuit.

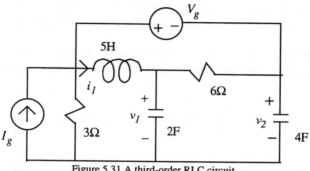

Figure 5.31 A third-order RLC circuit

5.2-3. Using the quantities identified for the circuit shown in **Figure 5.32** as the state variables, find the state equations of the circuit.

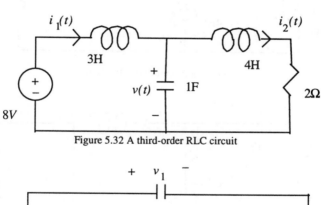

Figure 5.32 A third-order RLC circuit

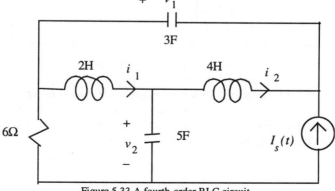

Figure 5.33 A fourth-order RLC circuit

5.2-4. Using the capacitor voltages and the inductor currents for the circuit shown in **Figure 5.33** as state variables, find the state equations of the circuit.

5.2-5. Using the capacitor voltages and the inductor currents for the circuit shown in **Figure 5.34** as state variables, find the state equations of the circuit.

5.2-6. Using the capacitor voltages and the inductor currents for the circuit shown in **Figure 5.35** as state variables, find the state equations of the following circuit.

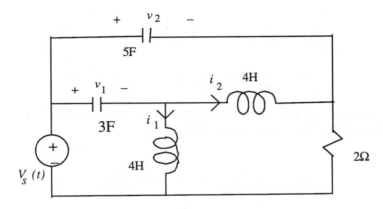

Figure 5.34 A fourth-order RLC circuit

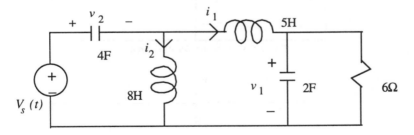

Figure 5.35 A fourth-order RLC circuit

5.3-1. The switch of the circuit shown in **Figure 5.36** has been opened for a long time and is closed at $t = 0$. Using the capacitor voltage $v(t)$ and the inductor current $i(t)$ as state variables, find the state equations of the circuit for $t > 0$, the roots of the characteristics equations, and the solution of the state equations.

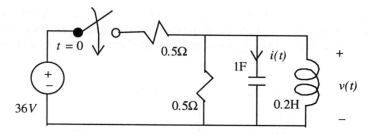

Figure 5.36 A second-order RLC circuit

5.3-2. The switch of the circuit shown in **Figure 5.37** has been at point '*a*' for a long time and is switched to point '*b*' at $t = 0$ Using the capacitor voltage $v(t)$ and the inductor current $i(t)$ as state variables, find the state equations of the circuit for $t > 0$ and the solution of the state equations.

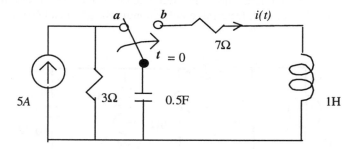

Figure 5.37 A second-order RLC circuit

5.3-3. The switch of the circuit shown in **Figure 5.38** has been closed for a long time and is opened at $t = 0$. Find the characteristic equation and its roots. Find the initial conditions $v(0+)$ and $v'(0+)$.

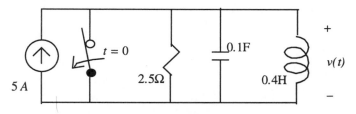

Figure 5.38 A second-order RLC circuit

5.3.4. The switch of the circuit shown in **Figure 5.39** has been closed for a long time and is opened at $t = 0$. Find the differential equation that governs $i(t)$ for $t > 0$, the roots of the characteristic equation, and the initial conditions $i(0+)$ and $i'(0+)$.

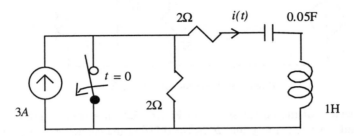

Figure 5.39 A second-order RLC circuit

5.3.5. The switch of the circuit shown in **Figure 5.40** has been open for a long time. At $t = 0$, the switch is closed. Determine initial conditions $i(0+)$ and $i'(0+)$.

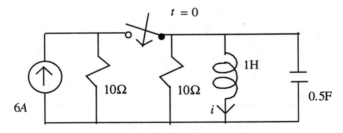

Figure 5.40 A second-order RLC circuit

Chapter 6
Sinusoidal Steady-State Analysis

When a circuit contains only sinusoidal sources of the same frequency, the circuit is often referred to as an AC circuit. For such a circuit, any circuit quantity is a sinusoidal function with the same frequency as the sources in the circuit. The process to find a circuit quantity in an AC circuit is known as sinusoidal steady-state analysis or AC analysis.

6.1 Sinusoids and Phasors

In sinusoidal steady-state analysis, sinusoidal circuit quantities (voltages or currents) with the same frequency are represented by complex numbers known as phasors. This phasor representation is based on the well-known Euler's Formula.

6.1.1 Euler's Formula and Complex Forcing Function

The famous Euler's Formula describes how a complex exponential function is related to the cosine and the sine functions as follows:

$$e^{j\theta} = \cos\theta + j\sin\theta \qquad 6.1\text{-}1$$

where j, known as the imaginary number, is the square root of -1 and the angle θ is measured in radians. The real part of the complex exponential function is the cosine function and the imaginary part is the sine function. When the angle is negative, we have

$$e^{-j\theta} = \cos(-\theta) + j\sin(-\theta) = \cos\theta - j\sin\theta \qquad 6.1\text{-}2$$

With these two formulas, the cosine function and the sine function can be expressed as linear combinations of complex exponential functions. When **Equations 6.1-1** and **6.1-2** are added together, we have

$$e^{j\theta} + e^{-j\theta} = 2\cos\theta$$

When **Equation 6.1-2** is subtracted from **Equation 6.1-1**, we have

$$e^{j\theta} - e^{-j\theta} = 2j\sin\theta$$

We thus have

$$\cos\theta = \frac{e^{j\theta} + e^{-j\theta}}{2} = \mathrm{Re}\{e^{j\theta}\} \qquad 6.1\text{-}3$$

and

$$\sin\theta = \frac{e^{j\theta} - e^{-j\theta}}{2j} = \mathrm{Im}\{e^{j\theta}\} \qquad 6.1\text{-}4$$

Since the angle θ can take any form, it can be a time function as well. In particular, we have

$$\cos(\omega \cdot t + \phi) = \frac{e^{j(\omega \cdot t + \phi)} + e^{-j(\omega \cdot t + \phi)}}{2}$$

and

$$\sin(\omega \cdot t + \phi) = \frac{e^{j(\omega \cdot t + \phi)} - e^{-j(\omega \cdot t + \phi)}}{2j}$$

Hence, a sinusoidal function can be considered as a linear combination of two complex exponential functions.

6.1.2 Sinusoids and Phasor Representation

When a signal takes the form of a cosine function, i.e.,

$$v(t) = V_m \cos(\omega \cdot t + \phi) \qquad 6.1\text{-}5$$

the signal is said to be a sinusoid. As shown in the form of a cosine function, a sinusoid can be properly identified with the triple $\{V_m, \omega, \phi\}$ which represent the amplitude, the frequency, and the phase angle, respectively. With the frequency specified, a sinusoid can be identified with the amplitude and the phase angle. Furthermore, we have

$$v(t) = V_m \cos(\omega \cdot t + \phi) = \mathrm{Re}\{V_m e^{j(\omega \cdot t + \phi)}\} = \mathrm{Re}\{V_m e^{j\phi} \cdot e^{j\omega \cdot t}\}$$

Consequently, the sinusoid with a given frequency can be identified with a complex number as follows:

$$V = (k_o V_m)e^{j\phi} = (k_o V_m)\angle^\phi \qquad 6.1\text{-}6$$

This complex number is known as the phasor representation of the sinusoid. In the phasor representation, k_o is a specific positive constant and is customarily chosen either as 1 or $(2)^{-1}$.

6.1.3 Particular Solutions to Linear Differential Equations

In **Figure 5.1** of the previous chapter, a first order RC circuit is shown. The voltage of the capacitor in the circuit is governed by the first order differential equation

$$\frac{dv}{dt} = \frac{i}{C} = \frac{1}{C}\left(\frac{v_s - v}{R}\right) = -\frac{v}{CR} + \frac{v_s}{CR} \qquad 6.1\text{-}7$$

The solution of the differential equation consists of two parts: the homogeneous solution and the particular solution. The homogenous solution takes the form of a exponentially decaying function of time t, i.e. $Ae^{-t/RC}$. The particular solution is assumed of the same form as the source function v_s. When v_s is a sinusoid, say $V_o\cos(\omega t + \delta)$, the particular solution is assumed to be $K\cos(\omega t + \gamma)$. Substituting the particular solution in **Equation 6.1-7**, we have

$$-\omega \cdot K \sin(\omega \cdot t + \gamma) = -\frac{K \cos(\omega \cdot t + \gamma)}{RC} + \frac{V_o \cos(\omega \cdot t + \delta)}{RC}$$

Solving for K and γ, we have

$$K = \frac{V_o}{\sqrt{1 + (\omega RC)^2}}$$

$$\gamma = \delta - \tan^{-1}(\omega RC)$$

As time increases, the homogeneous solution diminishes while the particular solution remains in effect. Consequently, the homogeneous solution is often referred to as the transient while the particular solution the steady-state response. Since the steady-state response to a sinusoidal source function remains a sinusoidal function of the same frequency, the amplitude and the phase angle of the sinusoidal response can be found via the so-called phasor technique instead of via solving the differential equation directly.

6.2 Phasor Technique

Once the voltage signals and current signals of the same frequency in a circuit are identified with phasors, the voltage-current characteristics of circuit elements are transformed into algebraic equations. These algebraic equations lead to the definition of the so-called impedance of circuit elements. With phasor representation and impedance, the steady-state analysis can be carried out through solving algebra equations instead of through solving differential equations.

6.2.1 Addition of Sinusoids, KCL, and KVL

In trigonometry, it is found that the sum of sinusoids of the same frequency remains a sinusoid with the same frequency. With the phasor representation, addition of sinusoids of the same frequency can be found through the addition of complex numbers.

We thus have

$$
\begin{aligned}
V_1 \cos(\omega \cdot t + \delta_1) + V_2 \cos(\omega \cdot t + \delta_2) &= \mathrm{Re}\{V_1 e^{j\delta_1} e^{j\omega \cdot t}\} + \mathrm{Re}\{V_2 e^{j\delta_2} e^{j\omega \cdot t}\} \\
&= \mathrm{Re}\{V_1 e^{j\delta_1} e^{j\omega \cdot t} + V_2 e^{j\delta_2} e^{j\omega \cdot t}\} = \mathrm{Re}\{[V_1 e^{j\delta_1} + V_2 e^{j\delta_2}] e^{j\omega \cdot t}\} \\
&= \mathrm{Re}\{V_0 e^{j\delta_0} e^{j\omega \cdot t}\} = V_0 \cos(\omega \cdot t + \delta_0)
\end{aligned}
$$

where

$$
\begin{aligned}
V_0 e^{j\delta_0} &= V_1 e^{j\delta_1} + V_2 e^{j\delta_2} \\
&= [V_1 \cos\delta_1 + V_2 \cos\delta_2] + j[V_1 \sin\delta_1 + V_2 \sin\delta_2]
\end{aligned}
$$

with

$$
\begin{aligned}
V_0 &= \{[V_1 \cos\delta_1 + V_2 \cos\delta_2]^2 + [V_1 \sin\delta_1 + V_2 \sin\delta_2]^2\}^{1/2} \\
&= \{V_1^2 + V_2^2 + 2V_1 V_2 \cos(\delta_1 - \delta_2)\}^{1/2}
\end{aligned}
$$

and

$$\delta_0 = \begin{cases} \tan^{-1} \dfrac{V_1 \sin \delta_1 + V_2 \sin \delta_2}{V_1 \cos \delta_1 + V_2 \cos \delta_2}, & if \quad V_1 \cos \delta_1 + V_2 \cos \delta_2 > 0 \\[3mm] 180° + \tan^{-1} \dfrac{V_1 \sin \delta_1 + V_2 \sin \delta_2}{V_1 \cos \delta_1 + V_2 \cos \delta_2}, & if \quad V_1 \cos \delta_1 + V_2 \cos \delta_2 < 0 \\[3mm] \pm 90°, & if \quad V_1 \cos \delta_1 + V_2 \cos \delta_2 = 0 \end{cases}$$

The most important application of the addition of phasors is found in KCL and KVL. For example, consider KCL. The law says that the algebraic sum of all current entering a node equals to zero. For the node shown in **Figure 2.1** of **Chapter 2**, we have

$$i_1 + i_2 - i_3 + i_4 - i_5 = 0$$

These currents can be time functions in general. In particular, they can be sinusoidal. Let's assume that they are sinusoids of the same frequency. We thus have

$$\begin{aligned} & i_1 + i_2 - i_3 + i_4 - i_5 \\ &= I_1 \cos(\omega \cdot t + \delta_1) + I_2 \cos(\omega \cdot t + \delta_2) - I_3 \cos(\omega \cdot t + \delta_3) \\ &\quad + I_4 \cos(\omega \cdot t + \delta_4) - I_5 \cos(\omega \cdot t + \delta_5) \\ &= \mathrm{Re}\{I_1 e^{j\delta_1} \cdot e^{j\omega \cdot t}\} + \mathrm{Re}\{I_2 e^{j\delta_2} \cdot e^{j\omega \cdot t}\} - \mathrm{Re}\{I_3 e^{j\delta_3} \cdot e^{j\omega \cdot t}\} \\ &\quad + \mathrm{Re}\{I_4 e^{j\delta_4} \cdot e^{j\omega \cdot t}\} - \mathrm{Re}\{I_5 e^{j\delta_5} \cdot e^{j\omega \cdot t}\} \\ &= \mathrm{Re}\{[I_1 e^{j\delta_1} + I_2 e^{j\delta_2} - I_3 e^{j\delta_3} + I_4 e^{j\delta_4} - I_5 e^{j\delta_5}] e^{j\omega \cdot t}\} \\ &= 0 \end{aligned}$$

The equality shown is true if and only if

$$\begin{aligned} & k_o(I_1 e^{j\delta_1} + I_2 e^{j\delta_2} - I_3 e^{j\delta_3} + I_4 e^{j\delta_4} - I_5 e^{j\delta_5}) \\ &= I_1 + I_2 - I_3 + I_4 - I_5 \\ &= 0 \end{aligned}$$

where I_m, with m ranging from 1 through 5, is the phasor representation of i_m as defined in **Equation 6.1-6**.

Similarly, when KVL is applied to the loop in **Figure 2.2** of **Chapter 2**, we have

$$v_1 + v_2 - v_3 + v_4 = 0$$

If these voltages are sinusoidal time functions with the same frequency, we conclude that

$$V_1 + V_2 - V_3 + V_4 = 0$$

where V_n, with n ranging from 1 through 4, is the phasor representation of v_n.

6.2.2 Differentiation of Sinusoids and Concept of Impedance

In calculus, it is found that the derivative of a sinusoid remains a sinusoid of the same frequency. Hence, the operation of differentiation of a sinusoid can be considered as a transformation of the sinusoid to another sinusoid of the same frequency but with different amplitude and phase angle. Consider the characteristics of an inductor as shown in **Figure 6.1(a)**.

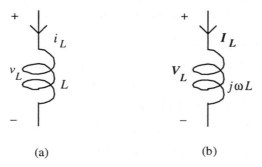

(a) (b)

Figure 6.1. An inductor and its phasor representation

Assuming that the current through the inductor is a sinusoid, we have

$$v_L = L\frac{di_L}{dt} = L\frac{d}{dt}[I_L \cos(\omega t + \delta)] = L[-I_L \omega \sin(\omega t + \delta)]$$
$$= I_L \omega L \cos(\omega t + \delta + 90^o)$$

Using phasor representation as defined in **Equation 6.1-6**, we have

$$I_L = k_o I_L \angle^{\delta}$$
$$V_L = k_o I_L \omega L \angle^{\delta + 90^o} = \omega L \angle^{90^o} \times k_o I_L \angle^{\delta} = j\omega L I_L$$

The equation above shows that the voltage phasor across an inductor is linearly proportional to the current phasor through the inductor. The proportional constant, $j\omega L$, is an imaginary number and can be assigned the same unit as that of the resistance of an resistor. This proportional constant is referred to as the impedance of the inductor. Identified with the impedance, an inductor can be represented by the symbol in **Figure 6.1(b)**.

Note that the impedance of an inductor is a linear function of the frequency. The characteristic of an inductor can further be demonstrated through the plotting of the voltage phasor and the current phasor in a two-dimensional complex plane as shown in **Figure 6.2**.

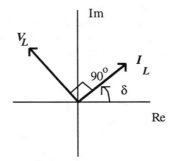

Figure 6.2. Phasor diagram for an inductor

Note that the phase angle of the current phasor is $90°$ less than the phase angle of the voltage phasor. Using the voltage phasor as a reference, the current phasor is said to be *lagging* the voltage phasor by $90°$.

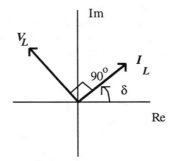

Figure 6.3. A capacitor and its phasor representation

Similarly for a capacitor as shown in **Figure 6.3(a)**, we have

$$i_C = C \frac{dv_C}{dt} = C \frac{d}{dt}[V_C \cos(\omega t + \phi)] = C[-V_C \omega \sin(\omega t + \phi)]$$

$$= V_C \omega C \cos(\omega t + \phi + 90^o)$$

With phasor representation as defined in **Equation 6.1-6**, we have

$$V_C = k_o V_C \angle^\phi$$

$$I_C = k_o V_C \omega C \angle^{\phi + 90^o} = \omega C \angle^{90^o} \times k_o V_C \angle^\phi = j\omega C V_C$$

The impedance of a capacitor, defined as the ratio of the voltage phasor across an element and the current phasor through the element, is thus $1/(j\omega C)$. The phasor diagram of a capacitor is shown in **Figure 6.4**. For a capacitor, the phase angle of the current phasor is 90° more than that of the voltage phasor. The current phasor is said to be *leading* the voltage phasor by 90°.

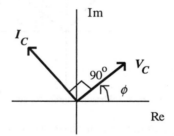

Figure 6.4. Phasor diagram for a capacitor

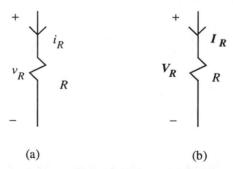

(a)   (b)

Figure 6.5. A resistor and its phasor representation

In the case of a resistor as in **Figure 6.5**(a), it can easily be shown that the impedance of the resistor is simply the resistance of the resistor, R. The phasor diagram of a resistor is shown in **Figure 6.6**. For a resistor, the phase angle of the current phasor is equal to that of the voltage phasor. The current phasor is said to be *aligned* with the voltage phasor.

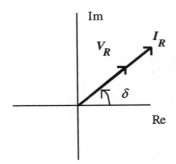

Figure 6.6. Phasor diagram for a resistor

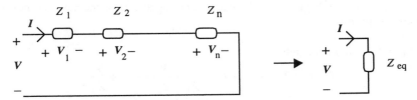

Figure 6.7. An one-port with elements in series and its equivalent circuit

6.2.3 Equivalent Circuits

When a group of elements (or one-ports) are connected in series to form a new one-port network as illustrated in **Figure 6.7**, the equivalent impedance of the new one-port can be found as follows:

We have

$$V = V_1 + V_2 + \cdots\cdots + V_n = IZ_1 + IZ_2 + \cdots\cdots + IZ_n$$
$$= I(Z_1 + Z_2 + \cdots\cdots + Z_n) = IZ_{eq}$$

Hence,

$$Z_{eq} = Z_1 + Z_2 + \cdots\cdots + Z_n$$

Furthermore,

$$V_k = I \cdot Z_k = \frac{V}{Z_1 + Z_2 + \ldots + Z_n} \cdot Z_k = \frac{Z_k}{Z_1 + Z_2 + \ldots + Z_n} \cdot V$$

where k varies from 1 to n.

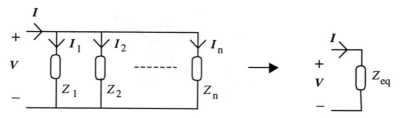

Figure 6.8. An one-port with elements in parallel and its equivalent circuit

When a group of resistors are connected in parallel to form a new one-port network as shown in **Figure 6.8**, the equivalent impedance of the new one-port can be found as follows:

From the circuit, we get

$$I = I_1 + I_2 + \cdots + I_n = \frac{V}{Z_1} + \frac{V}{Z_2} + \cdots + \frac{V}{Z_n}$$

$$= V(\frac{1}{Z_1} + \frac{1}{Z_2} + \cdots + \frac{1}{Z_n}) = \frac{V}{Z_{eq}}$$

Hence,

$$\frac{1}{Z_{eq}} = \frac{1}{Z_1} + \frac{1}{Z_2} + \cdots + \frac{1}{Z_n}$$

or, equivalently,

$$Z_{eq} = \left(\frac{1}{Z_1} + \frac{1}{Z_2} + \cdots + \frac{1}{Z_n} \right)^{-1}$$

Furthermore,

$$I_k = \frac{V}{Z_k} = \frac{I}{\frac{1}{Z_1} + \frac{1}{Z_2} + \ldots + \frac{1}{Z_n}} \cdot \frac{1}{Z_k} = \frac{\frac{1}{Z_k}}{\frac{1}{Z_1} + \frac{1}{Z_2} + \ldots + \frac{1}{Z_n}} \cdot I$$

where k varies from 1 to n.

Using series and parallel combinations, we can find the equivalent impedance of a certain kind of one-port network. This procedure can be demonstrated by solving for the current out of the voltage source in the circuit of **Figure 6.9**.

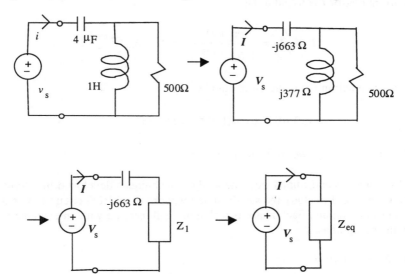

Figure 6.9. A one-port network connected to a sinusoidal source

The voltage source is assumed to be sinusoidal and with a frequency of 60 Hz as shown in the following equation:

$$v_s(t) = 150\cos(120\pi \cdot t + 90°)\ V$$

The phasor representation of this sinusoid is therefore

$$V_s = k_o 150\angle 90° = jk_o 150$$

The equivalent impedance Z_1 is found to be

$$Z_1 = \frac{1}{\dfrac{1}{j377} + \dfrac{1}{500}} = 301\angle 52.99°$$

The equivalent impedance Z_{eq} is found to be

$$Z_{eq} = -j663 + 301\angle^{52.99°} = 460\angle^{-66.83°}$$

The current phasor I is found to be

$$I = \frac{V_s}{Z_{eq}} = \frac{jk_o150}{460\angle^{-66.83°}} = k_o 0.326\angle^{156.83°}$$

The current function $i(t)$ is therefore equal to

$$i(t) = 0.326\cos(120\pi \cdot t + 156.83°) \qquad A$$

6.2.4 Analysis Techniques and Circuit Theorems

With phasors and impedance, the analysis techniques developed in **Chapter 2** can be generalized to study the steady-state response of an RLC circuit (AC circuit) with sinusoidal inputs. Each analysis technique is illustrated with an example in the following paragraphs.

6.2.4.1 Nodal Analysis

The nodal analysis is utilized to study the circuit in the following example.

Example 6.1 Find the current I in **Figure 6.10**.

Solution: Using the node voltage assigned in **Figure 6.10**, we have

$$\frac{V_1 - V_3}{50} + \frac{V_1 - V_2}{j200} + \frac{V_1}{-j120} = 0$$

$$\frac{V_2 - V_3}{50} + \frac{V_2 - V_1}{j\,200} + \frac{V_2}{60} = 0$$

Solving for V_1 and V_2, we find that

$$V_1 = 70.4 - j21.4$$

$$V_2 = 38.6 - j4.3$$

$$V_3 = 75 \angle 0°$$

For the current I, we find that

$$I = \frac{V_1 - V_2}{j200} = 0.181 \angle -118.62°$$

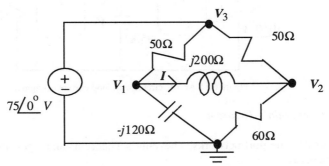

Figure 6.10 An AC circuit with node voltages assigned

6.2.4.2 Loop Analysis

The loop analysis is demonstrated with the following example.

Example 6.2 Find the voltage V of the circuit shown in **Figure 6.11**.

Solution: Using the loop currents assigned, we have

$$50I_1 + j100I_1 + (-j50)(I_1 - I_2) = 10 \angle 0°$$

$$j100I_2 + 50I_2 + (-j50)(I_2 - I_1) = 0$$

Solving for I_1 and I_2, we have

$$I_1 = 0.1264 \angle -18.4°$$

$$I_2 = 0.0894 \angle -153.4°$$

The voltage V is found to be

$$V = (-j50)(I_1 - I_2) = 4.89\angle^{-144.8^o} \quad volt$$

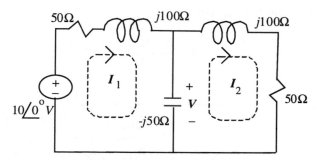

Figure 6.11 An AC circuit with loop currents assigned

6.2.4.3 Thevenin's Theorem

For AC one-port networks, Thevenin's Theorem can be generalized as shown in **Figure 6.12**.

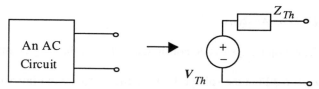

Figure 6.12 An AC one-port and its Thevenin's equivalent circuit

The generalized Thevenin's Theorem states that an AC one-port is equivalent to a simple one-port consisting of a voltage source and an impedance in series. In the equivalent circuit, the voltage source V_{Th} is the open-circuited voltage (in phasor form) across the two terminals. The impedance Z_{Th} is the equivalent impedance of the AC one-port with all internal independent sources set to zero. The usefulness of the theorem is demonstrated in the following example.

Example 6.3 Find the current I in **Figure 6.13** when (i) $Z_o = j75$ Ω and when (ii) $Z_o = 20 - j40$ Ω, respectively.

Solution: To find the Thevenin's equivalent impedance, we set all internal independent sources to zero as shown in **Figure 6.14**. The equivalent impedance of the one-port is found to be

$$Z_{Th} = 100 + \frac{1}{\frac{1}{10} + \frac{1}{(-j100)}} = 109.9 - j0.99$$

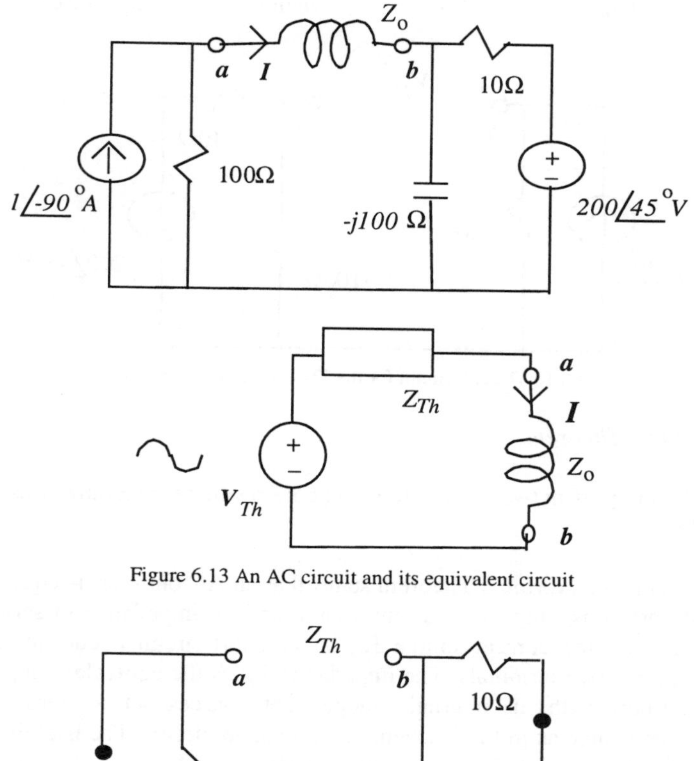

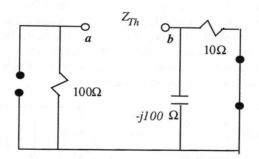

Figure 6.13 An AC circuit and its equivalent circuit

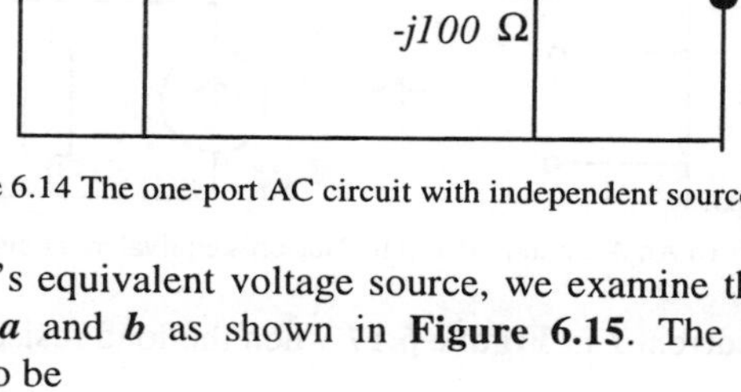

Figure 6.14 The one-port AC circuit with independent sources set to zero

For the Thevenin's equivalent voltage source, we examine the open-circuit voltage across terminals *a* and *b* as shown in **Figure 6.15**. The Thevenin's equivalent voltage is found to be

$$V_{Th} = 1\angle^{-90°} \times 100 - 200\angle^{45°} * \frac{-j100}{10 - j100} = -154 - j226$$

136

With the Thevinin's equivalent circuit, the current I can be found as follows:

$$I = \frac{V_{Th}}{Z_{Th} + Z_o} = \begin{cases} 2.06\angle^{201.8^o} & \text{when} & Z_o = j75\Omega \\ 2.01\angle^{253.2^o} & \text{when} & Z_o = 20 - j40\Omega \end{cases}$$

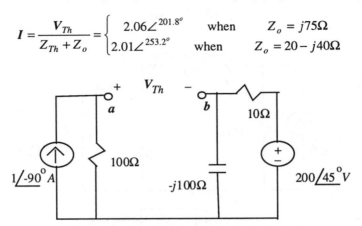

Figure 6.15 The AC circuit for the Thevinin's equivalent voltage

6.2.4.4 Norton's Theorem

For AC one-port networks, Norton's Theorem can be generalized as shown in **Figure 6.16**.

The generalized Norton's Theorem states that an AC one-port is equivalent to a simple one-port consisting of a current source and an impedance in series. In the equivalent circuit, the current source I_{Sh} is the short-circuited current (in phasor form) through the two terminals. The impedance Z_{Th} is the equivalent impedance of the AC one-port with all internal independent sources set to zero, the same equivalent impedance as in the Thevinin's equivalent circuit. The usefulness of the theorem is demonstrated in the following example.

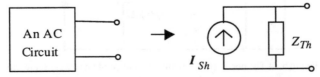

Figure 6.16 An AC one-port and its Norton's equivalent circuit

Example 6.4 Find the current I in **Figure 6.17** when the load resistance R is equal to 8Ω.

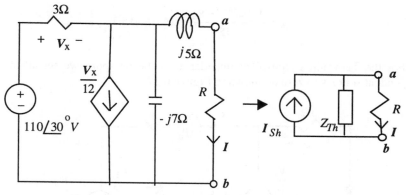

Figure 6.17 An AC circuit and its Norton's equivalent circuit

Solution: For the short-circuited current in the Norton's equivalent circuit, we short terminals **a** and **b** together and find the current through them as shown in **Figure 6.18**. Applying KCL at Node **c**, we have

$$\frac{110\angle 30^{o} - V_x}{j5} + \frac{110\angle 30^{o} - V_x}{-j7} + \frac{V_x}{12} - \frac{V_x}{3} = 0$$

The voltage V_x is found to be

$$V_x = 24.5\angle^{-47.12^{o}}$$

Figure 6.18 The one-port in Figure 6.17 with terminals short-circuited

The short-circuit current I_{Sh} in **Figure 6.18** is found to be

$$I_{Sh} = \frac{110\angle 30^{\circ} - V_x}{j5} = 14.6 - j15.7$$

For the Thevenin's equivalent impedance of the one-port, we set all internal independent sources to zero as shown in **Figure 6.19**.

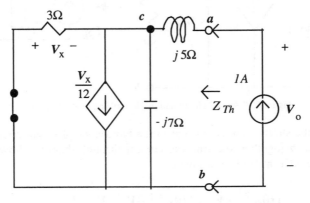

Figure 6.19 The one-port in Figure 6.17 with independent sources set to zero

To help finding the equivalent impedance, an auxiliary current source of $1A$ is connected between the terminals. Applying KCL to **Node c**, we have

$$\frac{V_x}{12} + \frac{V_x}{-j7} = \frac{V_x}{3} + 1$$

The voltage V_x is found to be

$$V_x = 3.47\angle^{-150.3^{\circ}}$$

The equivalent impedance Z_{Th} is found to be

$$Z_{Th} = \frac{V_o}{1} = V_o = j5 \times 1 - V_x = 3.01 + j6.72$$

With the Norton's equivalent circuit, the current I is found to be

$$I = \frac{Z_{Th}}{Z_{Th} + R} \times I_{Sh} = 12.22\angle^{-12.61^{\circ}}$$

6.3. Power Calculation

In **Chapter 1**, we learned that the power absorbed by an element in any circuit is equal to the product of the voltage across the element and the current through the element with the passive sign convention being observed. In a DC circuit, the value of the power absorbed by an element is constant since both the voltage and the current are constant in value. In an AC circuit, however, voltage signals and current signals are sinusoidal time functions of the same frequency. Consequently, the power absorbed by an element in an AC circuit is a varying time function. Fortunately, the power function is periodic in time. The average value of the power function can be found and is widely used in the power industry.

6.3.1 Average Power Consumption

The power absorbed by an element or a one-port network is generally a time function. This power is often referred to as the instantaneous power. For an AC circuit, the power function is periodic. The average value of the power function offers a good measure of the amount of power involved. The average value of the instantaneous power is called the average power. Since the average power is a real number, it is also referred to as the real power. Let's consider the circuit in **Figure 6.20**.

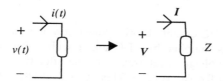

Figure 6.20. A one-port AC network and its phasor representation

For an AC circuit, both the voltage and the current are sinusoids of the same frequency. These two signals can be represented as follows:

$$v(t) = V_m \cos(\omega \cdot t + \phi)$$
$$i(t) = I_m \cos(\omega \cdot t + \delta)$$

The power absorbed by the one-port is then equal to

$$p(t) = v(t) \cdot i(t) = V_m I_m \cos(\omega \cdot t + \phi) \cos(\omega \cdot t + \delta)$$
$$= V_m I_m (\tfrac{1}{2})\{\cos(2\omega \cdot t + \phi + \delta) + \cos(\phi - \delta)\}$$

There are two terms in the expression for the instantaneous power, a sinusoid and a constant. It is well-known that the average value of a sinusoid is zero. Therefore the average value of the instantaneous power is given as follows:

$$P = (\tfrac{1}{2})V_m I_m \cos(\phi - \delta) = (\tfrac{1}{2})V_m I_m \cos\theta \qquad 6.3\text{-}1$$

where $\theta = \phi - \delta$ represents the phase angle of the equivalent impedance of the AC one-port network.

For any RLC one-port AC network containing no sources, the average power consumption is always greater than or equal to zero. Some special cases are discussed in the following sections.

6.3.1.1 Average Power Consumption by a Resistor

For a resistor with resistance R, the voltage signal and the current signal have the same phase angle. From **Equation 6.3-1**, the average power absorbed by a resistor is found to be

$$P_R = (\tfrac{1}{2})V_m I_m \cos(0^o) = (\tfrac{1}{2})V_m I_m = \frac{V_m^2}{2R} = \frac{I_m^2 R}{2}$$

Since a resistor in a DC circuit absorbs power at a constant value of $(V_{DC})^2/R$, the same power P_R can be supplied by a DC source with a terminal voltage of $V_m(2)^{-1/2}$. This equivalent DC voltage is referred to as the effective value or the root-mean-square (*rms*) value of the AC voltage. This equivalent DC voltage is denoted as V_{eff} or V_{rms}. In a similar fashion, an AC current is equivalent to a DC current with a value of $I_m(2)^{-1/2}$, denoted as I_{eff} or I_{rms}. With the effective values, **Equation 6.3-1** becomes the following:

$$P = V_{rms} I_{rms} \cos(\phi - \delta) = V_{rms} I_{rms} \cos\theta \qquad 6.3\text{-}2$$

Equation 6.3-2 is widely used in power calculations for AC circuits. To further facilitate the usage of **Equation 6.2**, the phasor representation of AC voltages and currents is often done with the effective values. Namely, the constant k_o is chosen to be $\sqrt{2}$ as follows:

$$V = k_o V_m \angle^\phi = \sqrt{2} \cdot V_m \angle^\phi$$

$$I = k_o I_m \angle^\delta = \sqrt{2} \cdot I_m \angle^\delta$$

From this point on, all voltage phasors and current phasors are assumed to be in *rms*-value.

6.3.1.2 Average Power Consumption by an Inductor or a Capacitor

For an inductor, the phase angle difference between the voltage and the current is 90°. From **Equation 6.3-2**, the average power absorbed by the inductor is found to be zero as shown in the following:

$$P_L = V_{rms} I_{rms} \cos(90°) = 0$$

In the case of a capacitor, the phase angle difference between the voltage and the current is -90°. Therefore, the average power absorbed by the capacitor is also equal to zero as shown in the following:

$$P_C = V_{rms} I_{rms} \cos(-90°) = 0$$

6.3.1.3 Examples

The power absorbed by a passive one-port network can be found through its equivalent circuit or through power calculations for every element in the one-port network. Assuming there are m passive elements in an one-port network, the average power absorbed by the one-port is equal to the summation of individual powers absorbed by every element in the circuit as shown in the following equation:

$$P_{total} = \sum_{i=1}^{m} P_i = \sum_{i=1}^{m} V_i I_i \cos \theta_i \qquad \text{6.3-3}$$

In **Equation 6.3-3**, voltages and currents are assumed to be in *rms*-value.

Example 6.4 Find the power absorbed by the 1-port network shown in **Figure 6.21**.

Solution: Since there are two elements in the one-port, we have

$$P_{total} = P_R + P_C = I^2 R + 0 = I^2 R$$

Alternatively,

$$P_{total} = VI\cos\theta = (I\sqrt{R^2 + X_C^2})I\cos\theta = I^2(\sqrt{R^2 + X_C^2}\cos\theta) = I^2 R$$

where we have made use of the fact that

$$Z_s = R + jX_c = \sqrt{R^2 + X_c^2}\angle^{\tan^{-1}(X_c/R)} = \sqrt{R^2 + X_c^2}\angle\theta$$

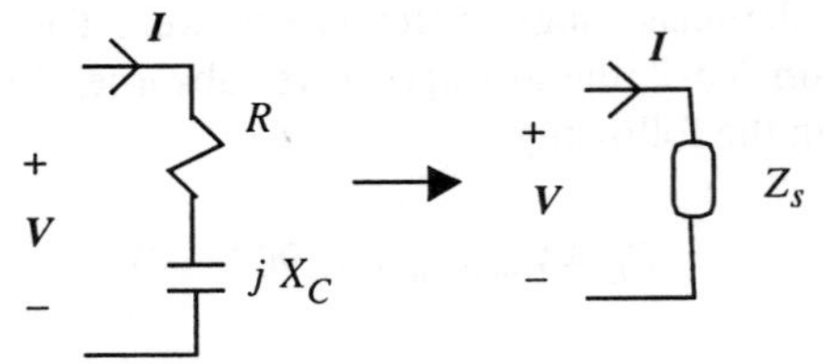

Figure 6.21. An one-port AC network with series elements and its equivalent

Example 6.5 Find the power absorbed by the 1-port network shown in **Figure 6.22**.

Solution: Since there are two elements in the one-port, we have

$$P_{total} = P_L + P_R = 0 + \frac{V^2}{R} = \frac{V^2}{R}$$

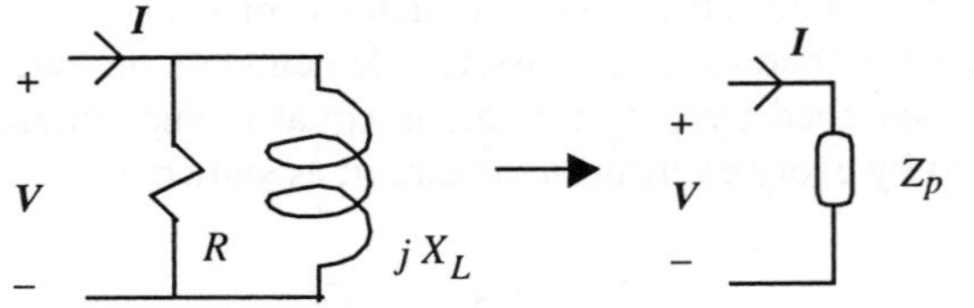

Figure 6.22. An one-port AC network with parallel elements and its equivalent

Alternatively, we have

$$Z_p = \frac{1}{\dfrac{1}{R} + \dfrac{1}{jX_L}} = \frac{1}{\dfrac{1}{R} - j\dfrac{1}{X_L}} = \frac{1}{\sqrt{R^{-2} + X_L^{-2}}\angle^{-\theta}} = \rho_p\angle\theta$$

where

$$\rho_p = \frac{1}{\sqrt{R^{-2} + X_L^{-2}}} \quad and \quad \theta = \tan^{-1}\left[\frac{1/X_L}{1/R}\right]$$

In addition,

$$I = V/Z_p = \frac{V\angle\phi}{\rho_p\angle\theta} = V\sqrt{R^{-2} + X_L^{-2}}\angle\phi-\theta$$

The average power absorbed by the equivalent impedance Z_p is thus found to be

$$P_{total} = VI\cos\theta$$

$$= V(V\sqrt{R^{-2} + X_L^{-2}})\cos\theta = V^2(\sqrt{R^{-2} + X_L^{-2}}\cos\theta) = \frac{V^2}{R}$$

where we have made use of the fact that

$$\cos\theta = \frac{1/R}{\sqrt{R^{-2} + X_L^{-2}}}$$

Example 6.6 Find the power absorbed by all passive elements of the circuit shown in **Figure 6.23**, when V_s is 110 V, *rms*.

Solution: For the circuit shown, the equivalent impedance is given as follows:

$$Z_{eq} = 2 + \frac{4 \times j4}{4 + j4} = 4 + j2$$

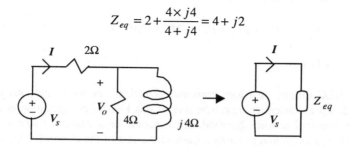

Figure 6.23. An AC network and its equivalent

The power absorbed by all the passive elements is found to be

$$P_{total} = VI \cos \theta = 110(\frac{110}{\sqrt{4^2+2^2}}) \frac{4}{\sqrt{4^2+2^2}} = 2420 \quad W$$

Alternatively, we have

$$P_{total} = P_{R1} + P_{R2} + P_L$$

The first term in the equation above is found to be

$$P_{R1} = I^2 R_1 = \left(\frac{110}{\sqrt{4^2+2^2}}\right)^2 * 2 = 1210 \ W$$

To calculate the second term, we first find the voltage V_o.

$$V_o = IZ_o = \frac{110}{\sqrt{4^2+2^2}} * \frac{1}{\sqrt{4^{-2}+4^{-2}}} = 220/\sqrt{10} \quad V$$

The second term is then found to be

$$P_{R2} = \frac{V_o{}^2}{R_2} = \frac{\left(220/\sqrt{10}\right)^2}{4} = 1210 \quad W$$

Hence, the total power is equal to

$$P_{total} = P_{R1} + P_{R2} + P_L = 1210 + 1210 + 0 = 2420 \quad W$$

6.3.2 Complex Power

In an RLC circuit, resistors consume electrical energy continuously without giving back any electrical energy. The rate of this kind of consumption is measured by the average real power absorbed. An inductor or a capacitor stores electrical energy for half a cycle and releases it back to the rest of the circuit for the other half a cycle, alternatingly. This type of energy transfer is not reflected in the value of the average power. To account for the energy transfer between the energy storage elements and the sources, a complex-valued quantity called complex power is thus defined.

The complex power absorbed by a one-port network shown in **Figure 6.20** is defined as follows:

$$S = V(I)^*$$

6.3-4

where the asterisk denotes the operation of taking the conjugate of a complex number.

With phasor representation, the complex power can be found as follows:

$$S = V(I)^* = V\angle^\phi \left(I\angle^\delta\right)^* = V\angle^\phi \times \left(I\angle^{-\delta}\right) = VI\angle^{(\phi-\delta)}$$
$$= VI\angle^\theta = VI\cos\theta + jVI\sin\theta = P + jQ$$

where $Q = VI\sin\theta$ is known as the reactive power.

The magnitude of the complex power is the product of the voltage *rms* value and the current *rms* value. It is called the apparent power and is measured in the unit of *VA* (*volt* times *amp*). The real part of the complex power represents the average power absorbed by the one-port network. The reactive power reflects the effect of energy storage elements, inductors or capacitors. A source connected to the one-port network must not only to supply the real power absorbed by the one-port but also to accommodate the energy transfer among the energy storage elements and the source itself. To emphasize the difference between the reactive power and the real power, the reactive power is measured in the units of *VAR*.

For a resistive one-port, no electrical energy is stored first and then released back to the rest of the circuit. Therefore, the reactive power is zero for this type of network. For an energy storage element, the reactive power can be found as demonstrated by the following examples.

Example 6.7 Find the complex power absorbed by a capacitor as shown in **Figure 6.24**.

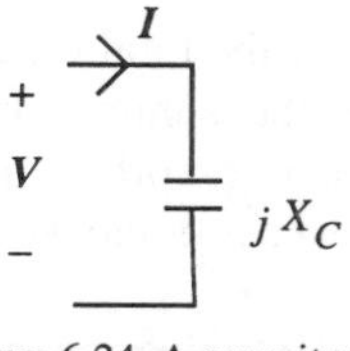

Figure 6.24. A capacitor

Solution: The complex power consumed by the capacitor is found to be

$$S_C = V(I)^* = (I \cdot jX_C)(I)^* = jX_C I^2 = jQ_C \qquad 6.3\text{-}5$$

The reactive power is equal to

$$Q_C = I^2 X_C = I^2 \left(-\frac{1}{\omega \cdot C}\right) = -\frac{I^2}{\omega \cdot C} = -\frac{1}{\omega \cdot C}(V\omega \cdot C)^2 = -V^2 \omega \cdot C = \frac{V^2}{X_C}$$

It should be noted that the reactive power consumed by a capacitor is negative in value as X_C is negative.

Example 6.8 Find the complex power absorbed by an inductor as shown in **Figure 6.25**.

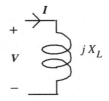

Figure 6.25. An inductor

Solution: The complex power consumed by the inductor is found to be

$$S_L = V(I)^* = (I \cdot jX_L)(I)^* = jX_L I^2 = jQ_L \qquad 6.3\text{-}6$$

The reactive power is equal to

$$Q_L = I^2 X_L = I^2 (\omega \cdot L) = \left(\frac{V}{\omega \cdot L}\right)^2 \omega \cdot L = \frac{V^2}{\omega \cdot L} = \frac{V^2}{X_L}$$

Similar to the total real power, the total reactive power consumed by a one-port passive network is found to be the summation of the individual reactive power consumed by every element in the one-port. So is the complex power. This property of reactive power or complex power is demonstrated with the following example.

Example 6.9 Find the total complex power consumed by all the passive elements of the circuit shown in **Figure 6.26**, when V_s is 10 V, *rms*.

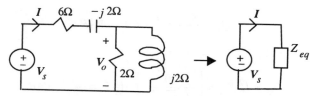

Figure 6.26. An AC network and its equivalent

Solution: For the circuit shown, the equivalent impedance is given as follows:

$$Z_{eq} = 6 - j2 + \frac{1}{\dfrac{1}{2} + \dfrac{1}{j2}} = 7 - j1$$

The source current is found to be

$$I = \frac{V_s}{Z_{eq}} = \frac{10\angle^\phi}{7 - j1}$$

The total complex power is equal to

$$S_{total} = V_s(I)^* = 10\angle^\phi \times \left(\frac{10e^{j\phi}}{7 - j1}\right)^* = 10\angle^\phi \times \left(\frac{10\angle^{-\phi}}{7 + j1}\right) = \frac{100}{7 + j1} = 14 - j2$$

Alternatively, we can find the power absorbed by every element first and then add the individual power to find the total power. Firstly, the voltage V_o is found to be

$$V_o = I \frac{1}{\dfrac{1}{2} + \dfrac{1}{j2}} = \frac{10\angle^\phi}{7 - j1} \times (1 + j) = \frac{(6 + j8)\angle^\phi}{5} = 2\angle^{\phi + 53.13^o}$$

We then have the following:

$$P_{2\Omega} = \frac{V_o^2}{2} = \frac{2^2}{2} = 2 \quad W$$

$$P_{6\Omega} = I^2 \cdot 6 = \left(\frac{10}{\sqrt{7^2 + 1^2}}\right)^2 \cdot 6 = 12 \quad W$$

$$Q_L = \frac{V_o^2}{X_L} = \frac{2^2}{2} = 2 \quad VAR$$

$$Q_C = I^2 X_C = \left(\frac{10}{\sqrt{7^2 + 1^2}}\right)^2 \cdot (-2) = -4 \quad VAR$$

Therefore,

$$Q_{total} = Q_L + Q_C = 2 + (-4) = -2 \quad VAR$$

$$P_{total} = P_{2\Omega} + P_{6\Omega} = 2 + 12 = 14 \quad W$$

Hence, the total complex power is equal to

$$S_{total} = P_{total} + jQ_{total} = 14 + j(-4) = 14 - j2 \quad VA$$

6.3.3 Power Factor and Power Factor Correction

There are three terms in **Equation 6.3-2** that determine the average power: the voltage *rms*-value, the current *rms*-value, and $\cos\theta$. The last term is referred to as the power factor (*pf*) which can be found as follows:

$$pf = \cos\theta = \frac{P}{V_{rms} I_{rms}} \qquad 6.3\text{-}7$$

The power factor of a particular passive AC one-port network is often identified either as lagging for positive θ or leading for negative θ, respectively. Power factor calculation can be illustrated with the following example.

Example 6.10 For the circuit shown in **Figure 6.27**, an AC load represented by the impedance Z_l consumes an average power of 76 *kW* at a lagging power factor of 0.731 when it is operating at 60 *Hz* and 440 *V*, *rms* value. The impedance Z_{tran} models the transmission lines utilized by the power source for power supply. The value of the impedance in the transmission lines is given at $0.09 + j0.11$ Ω. Find the average power that the voltage source must supply and the value of the source voltage.

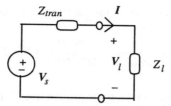

Figure 6.27. An equivalent AC circuit representing a power source and its load

Solution: At the operating voltage, the current in the load is found to be

$$I_l = \frac{P_l}{V_l \cos(\theta_l)} = \frac{76000}{440 \times 0.731} = 236.3 \quad A$$

The average power consumed in the transmission lines is equal to

$$P_{trans} = I_l^2 R_{trans} = (236.3)^2 \times 0.09 = 5025 \quad W$$

The total power needed is equal to

$$P_{total} = P_l + P_{trans} = 76000 + 5025 = 81025 \quad W$$

To find the voltage of the voltage source, we first find the complex powers involved. The complex power consumed in the transmission lines is equal to

$$S_{trans} = I_l^2 Z_{trans} = (236.3)^2 \times (0.09 + j0.11) = 5025 + j6142 \quad VA$$

The total complex power needed is equal to

$$\begin{aligned} S_{total} &= S_{trans} + S_l \\ &= (5025 + j6142) + (76000 + j76000 \tan\theta_l) = 81025 + j77087 \quad VA \end{aligned}$$

From the total complex power, the source voltage is found to be:

$$V_s = \frac{S_{total}}{I_l} = \frac{\sqrt{81025^2 + 77087^2}}{236.3} = 473.3 \quad V$$

Example 6.11 Repeat the calculations in **Example 6.10** if the power factor of the load is 0.92, lagging.

Solution: With the new power factor, the current in the load is found to be

$$I_l = \frac{P_l}{V_l \cos(\theta_l)} = \frac{76000}{440 \times 0.92} = 187.7 \quad A$$

The average power consumed in the transmission lines is reduced to

$$P_{trans} = I_l^2 R_{trans} = (187.7)^2 \times 0.09 = 3171 \quad W$$

The total power needed is also reduced to

$$P_{total} = P_l + P_{trans} = 76000 + 3171 = 79171 \quad W$$

The complex power consumed in the transmission lines is equal to

$$S_{trans} = I_l^2 Z_{trans} = (187.7)^2 \times (0.09 + j0.11) = 3171 + j3875 \quad VA$$

The total complex power needed is equal to

$$S_{total} = S_{trans} + S_l$$
$$= (3171 + j3875) + (76000 + j76000 \tan \theta_l) = 79171 + j36251 \quad VA$$

From the total complex power, the source voltage is found as follows:

$$V_s = \frac{S_{total}}{I_l} = \frac{\sqrt{79171^2 + 36251^2}}{187.7} = 463.9 \quad V$$

The results in **Example 6.11** demonstrated that the closer the power factor is to 1.0, the less the demands on the voltage source. To effectively changing the power factor of a load without changing the real power consumption of the load is the subject known as *power factor correction*. One way to improve the power factor is to connect some energy storage elements in parallel with a load with undesirable power factor. For a load with lagging power factor, capacitors are used to offset the positive reactive power consumption of the load. Such an arrangement as shown in **Figure 6.28** has two distictions: (1) no added real power consumption and (2) the ease of maintaining the same operating load voltage.

Let the total complex power consumed by the load and the capacitor together be denoted as S_d. This complex power is found to be

$$S_d = S_l + S_C = (P_l + jQ_l) + jQ_C = P_l + j(Q_l + Q_C) = S_d \angle \theta_d \quad VA$$

Furthermore, we have

$$\tan \theta_d = \frac{Q_l + Q_C}{P_l} = \frac{P_l \times \tan \theta_l + Q_C}{P_l}$$

Therefore, the reactive power consumed by the capacitor is equal to

$$Q_C = P_l \times \tan \theta_d - P_l \times \tan \theta_l = -\omega C V_l^2$$

The capacitance is then found to be

$$C = \frac{P_l \times \tan \theta_d - P_l \times \tan \theta_l}{-\omega V_l^2} = \frac{P_l \times \tan \theta_l - P_l \times \tan \theta_d}{\omega V_l^2}$$

For the kind of power factor improvement stated in **Example 6.11**, the capacitance needed is found to be

$$C = \frac{76000 \times \left[\tan\left(\cos^{-1} 0.731 \right) - \tan\left(\cos^{-1} 0.92 \right) \right]}{2\pi \times 60 \times 440^2} = 528 \times 10^{-6} \, F = 528 \mu F$$

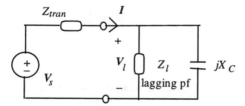

Figure 6.28. Power factor correction for a lagging load

Problems

6.1-1. Find the phasor representation for each of the following functions:

(a) $i(t) = -25 \sin (120\pi t + 123°) \, A$.
(b) $v(t) = 125 \cos (120\pi t - 47°) \, V$.

(c) $v(t) = 440 \sin(\omega t - 220°)$ V.

(d) $i(t) = -3.7 \cos(\omega t + 123°)$ A.

(e) $v(t) = -70 \cos(100t - 23°) + 80 \sin(100t+66°)$ V.

6.1-2. Find the phasor representation for the current $i(t)$ through the inductor, with the voltage $v(t) = 440 \cos(120\pi t-37°)$ V across the inductor as shown in **Figure 6.29**.

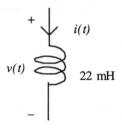

Figure 6.29. An inductor under AC

6.1-3. Find the phasor representation for the voltage, $v(t)$, across the capacitor with $i(t) = 20 \cos(120\pi t+50°)$ A as shown in **Figure 6.30**.

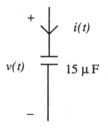

Figure 6.30. A capacitor under AC

6.2-1. Find the equivalent impedance Z (in rectangular or polar form) for each of the circuits shown in **Figure 6.31** with $\omega = 377$ radians per second.

6.2-2. Find the equivalent impedance of the circuit in **Figure 6.32**.

6.2-3. For the circuit shown in **Figure 6.33**, $V_s(t) = 15 \cos(2t + 33°)$ V. Find $i(t)$.

6.2-4. Using the node voltages assigned for the circuit shown in **Figure 6.34**, write down the network equations.

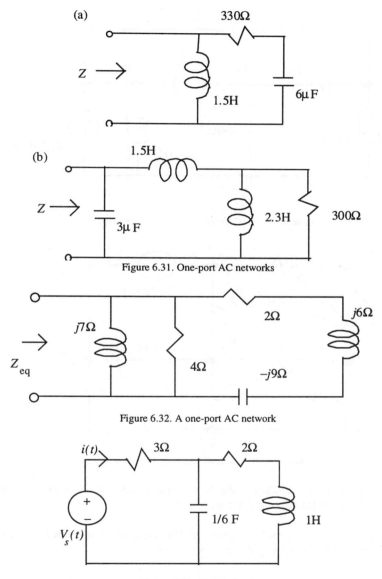

Figure 6.31. One-port AC networks

Figure 6.32. A one-port AC network

Figure 6.33. An AC circuit

6.2-5. Using the node voltages assigned in **Figure 6.35**, write down the network equations necessary for solving these node voltages.

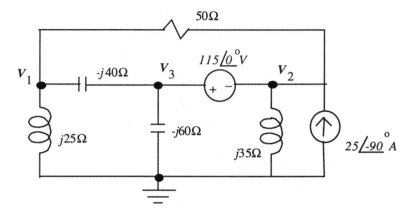

Figure 6.34. An AC circuit with node voltages assigned

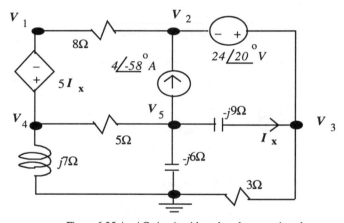

Figure 6.35 An AC circuit with node voltages assigned

6.2-6. Using the node voltages assigned in **Figure 6.36**, write down the network equations necessary for solving these node voltages.

6.2-7. Using the loop currents assigned in **Figure 6.37**, write down the network equations necessary for solving these loop currents.

6.2-8. Using the loop currents assigned in **Figure 6.38**, write down the network equations necessary for solving these loop currents.

6.2-9. Find the Thevenin's equivalent circuit of the circuit in **Figure 6.39**.

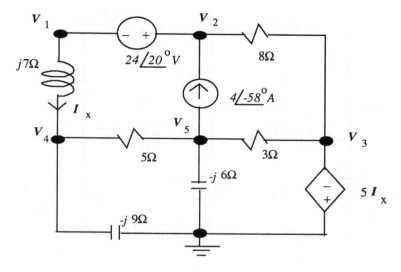

Figure 6.36 An AC circuit with node voltages assigned

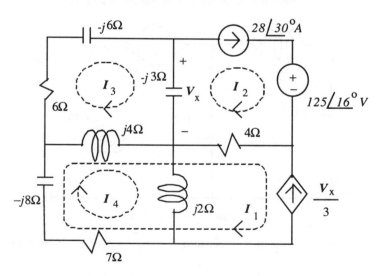

Figure 6.37 An AC circuit with loop currents assigned

6.2-10. Find the Norton equivalent circuit for the circuit in **Figure 6.40**.

6.3-1. Calculate the real power consumption by each of the circuits in **Figure 6.41**.

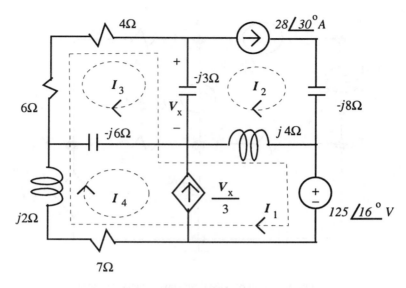

Figure 6.38 An AC circuit with loop currents assigned

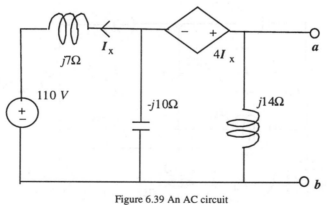

Figure 6.39 An AC circuit

6.3-2. Calculate the real power consumption by each of the following circuit in **Figure 6.42**.

6.3-3. The load impedance Z_L of the circuit shown in **Figure 6.43** is adjusted until maximum average power is delivered to Z_L. Find the value of Z_L and the maximum average power delivered to Z_L.

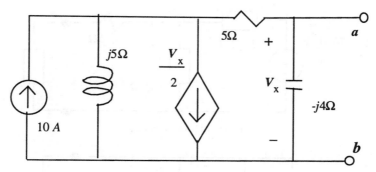

Figure 6.40 An AC circuit

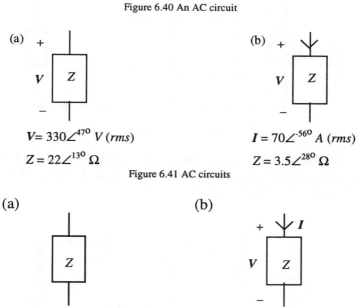

(a)

$V = 330\angle 47^0$ V (rms)

$Z = 22\angle 13^0$ Ω

(b)

$I = 70\angle -56^0$ A (rms)

$Z = 3.5\angle 28^0$ Ω

Figure 6.41 AC circuits

(a)

$|S| = 75$ kVA

$Z = 40 + j20$ Ω

(b)

$V = 75$ V (rms), $I = 23$ A (rms)

$Q = -1000$ VAR

Figure 6.42 AC circuits

6.3-4. The load impedance Z_L of the circuit shown in **Figure 6.44** is adjusted until maximum average power is delivered to Z_L. Find the value of Z_L.

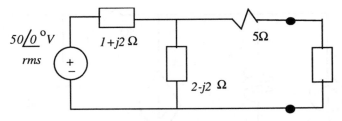

Figure 6.43 An AC circuit

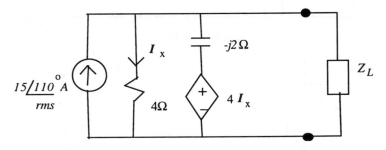

Figure 6.44 An AC circuit

6.3-5. The load resistor R_L of the circuit shown in **Figure 6.45** is adjusted until maximum average power is delivered to R_L. Find the value of R_L and the maximum average power delivered to R_L.

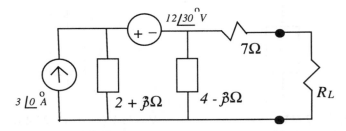

Figure 6.45 An AC circuit

6.3-6. A transmission line with impedance $0.06 + j0.13$ Ω is used by a power company to deliver power to two industrial loads as shown in **Figure 6.46**. Find the losses in the transmission line and the voltage V_s that the power company needs to generate.

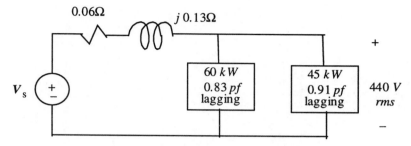

Figure 6.46 An AC source supplying power to two loads

Chapter 7
Three-Phase Circuits

Three-phase circuits are universally utilized by the power industry. First, a power company generates electricity with a three-phase synchronous generator. The generated electricity is then transmitted through three-phase power lines over a long distance to a power substation near a group of users. Finally, the transmitted electricity is then distributed to individual users by the substation. Three-phase power can be found in those factories where a large amount of power is needed to drive three-phase induction motors.

7.1 Three-Phase Sources

A three-phase source actually consists of three sinusoidal voltage sources of the same amplitude and the same frequency. In addition, the phase angles of any two of the three voltage sources differ by exactly 120 degrees. These three sources can be represented as follows:

$$v_{a1}(t) = V_m \cos(\omega_o t + \phi) \ V$$

$$v_{b2}(t) = V_m \cos(\omega_o t + \phi - 120^o) \ V$$

$$v_{c3}(t) = V_m \cos(\omega_o t + \phi + 120^o) \ V$$

Using the phasor technique, these three sources can also be identified with the following

$$V_{a1} = \frac{V_m}{\sqrt{2}} \angle^{\phi} \ V$$

$$V_{b2} = \frac{V_m}{\sqrt{2}} \angle^{\phi - 120^o} \ V$$

$$V_{c3} = \frac{V_m}{\sqrt{2}} \angle^{\phi + 120^o} \ V$$

These three sources are often referred to as the source phase voltages. Due to the fact that the magnitudes of these three phase voltages are of the same value, the magnitude of each source phase voltage is denoted as V_{ps}. Symbolically, the three sources can be represented as shown in **Figure 7.1**.

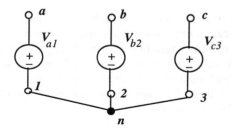

Figure 7.1 Three-phase sources in Y-connection

Note that Terminals *1*, *2*, and *3*, are tied together to form the node *n* known as the neutral. The three sources can also be plotted in a phasor diagram as shown in **Figure 7.2**. In this figure, the angle ϕ is assumed to be zero without losing any generality.

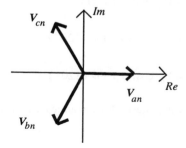

Figure 7.2 The phasor diagram of a three-phase source

Due to symmetry, we also have

$$V_{an} + V_{bn} + V_{cn} = 0$$

The voltages between terminal pairs *a* & *b*, *b* & *c*, or *c* & *a* are known as line voltages of the source. Line voltage V_{ab} can be found as follows:

$$V_{ab} = V_{an} - V_{bn} = V_{ps} \angle^{\phi} - V_{ps} \angle^{\phi - 120^{o}} = V_{ps} \angle^{\phi} \times [1 - 1 \angle^{-120^{o}}]$$

$$= V_{ps} \angle^{\phi} \times [1 - \cos(-120^o) - j\sin(-120^o)] = V_{ps} \angle^{\phi} \times [1 + \frac{1}{2} + j\frac{\sqrt{3}}{2}]$$

$$= V_{ps} \angle^{\phi} \times \sqrt{3} \angle^{30^o} = \sqrt{3} V_{ps} \angle^{\phi+30^o}$$

Similarly, we have

$$V_{bc} = V_{bn} - V_{cn} = V_{ps} \angle^{\phi-120^o} - V_{ps} \angle^{\phi+120^o} = \sqrt{3} V_{ps} \angle^{\phi-90^o}$$

$$V_{ca} = V_{cn} - V_{an} = V_{ps} \angle^{\phi+120^o} - V_{ps} \angle^{\phi} = \sqrt{3} V_{ps} \angle^{\phi+150^o}$$

Note that the magnitudes of these three line voltages are of the same value. The magnitude of each source line voltage is denoted as V_{Ls}. We have

$$V_{Ls} = \sqrt{3} V_{ps} \qquad\qquad 7.1\text{-}1$$

The difference in phase angle between a pair of line voltages is exactly 120°. The relative locations of line voltages with respect to phase voltages are illustrated in **Figure 7.3**.

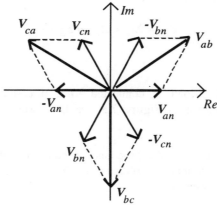

Figure 7.3 The phasor diagram of phase voltages and line voltages

7.2 Three-Phase Loads

When a three-phase source is available, loads are normally connected to the three-phase source in one of two ways: (a) a Y-load or (b) a Δ-load as shown in **Figure 7.4**.

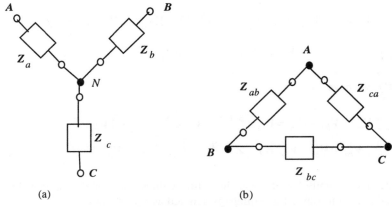

Figure 7.4 Three-phase loads

A Y-load is said to be a balanced load if

$$Z_a = Z_b = Z_c = Z_Y$$

A Δ-load is said to be a balanced load if

$$Z_{ab} = Z_{bc} = Z_{ca} = Z_\Delta$$

7.2.1 Balanced Y-Loads

When a balanced Y-load is connected to a three-phase source through transmission lines as shown in **Figure 7.5**, we have what is known as a Y-Y balanced power system.

Voltage V_{AN} is the voltage across the impedance Z_Y connected between Nodes A and N. This voltage is known as a phase voltage of the Y-load. If the impedance of each transmission line is zero, we have

$$V_{AN} = V_{an} = V_{ps} \angle^\phi$$

Similarly, we have

$$V_{BN} = V_{bn} = V_{ps} \angle^{\phi-120^o}$$

$$V_{CN} = V_{cn} = V_{ps} \angle^{\phi+120^o}$$

As noted before, the magnitudes of these three phase voltages are of the same value. The magnitude of each Y-load phase voltage is denoted as V_{pY}.

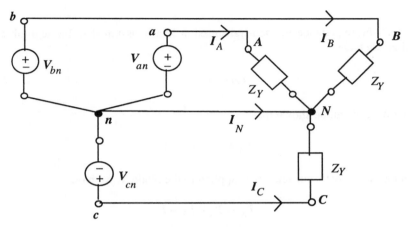

Figure 7.5 An Y-Y power system

Current I_A is the current flowing in one of the transmission lines. It is known as a line current of the Y-load. Since this current also flows in the impedance Z_Y, it is also referred to as a phase current of the Y-load. Similarly, currents I_B and I_C are known as line currents that are also phase currents of the Y-load. These currents can be found as follows:

$$I_A = \frac{V_{AN}}{Z_Y} = \frac{V_{an}}{\rho_Y \angle^{\theta_Y}} = \frac{V_{ps} \angle^{\phi}}{\rho_Y \angle^{\theta_Y}}$$

$$= \frac{V_{ps}}{\rho_Y} \angle^{\phi-\theta_Y} = I_{pY} \angle^{\phi-\theta_Y} = I_{LY} \angle^{\phi-\theta_Y}$$

$$I_B = \frac{V_{BN}}{Z_Y} = \frac{V_{bn}}{\rho_Y \angle^{\theta_Y}} = \frac{V_{ps} \angle^{\phi-120^o}}{\rho_Y \angle^{\theta_Y}}$$

$$= \frac{V_{ps}}{\rho_Y} \angle^{\phi-120^o-\theta_Y} = I_{pY} \angle^{\phi-120^o-\theta_Y} = I_{LY} \angle^{\phi-120^o-\theta_Y}$$

$$I_C = \frac{V_{CN}}{Z_Y} = \frac{V_{cn}}{\rho_Y \angle^{\theta_Y}} = \frac{V_{ps} \angle^{\phi+120^o}}{\rho_Y \angle^{\theta_Y}}$$

$$= \frac{V_{ps}}{\rho_Y} \angle^{\phi+120^o - \theta_Y} = I_{pY} \angle^{\phi+120^o - \theta_Y} = I_{LY} \angle^{\phi+120^o - \theta_Y}$$

Note that the impedance in each transmission line is assumed to be negligible. In addition, we have

$$I_{LY} = I_{pY} \qquad\qquad 7.2\text{-}1$$

Furthermore, it can be shown that

$$I_A + I_B + I_C = 0$$

When the Kirchhoff's current law is applied to the Node N, we have

$$-I_N = I_A + I_B + I_C$$

Hence, the current I_N flowing between Node n and Node N is zero. Consequently, the voltage between these two nodes is zero. For a balanced power system, the connection between Node n and Node N is not necessary.

The voltages between terminal pairs A & B, B & C, or C & A are known as line voltages of the load. Line voltage V_{AB} can be found as follows:

$$V_{AB} = V_{AN} - V_{BN} = V_{an} - V_{bn} = V_{ab} = \sqrt{3}\, V_{ps} \angle^{\phi+30^o}$$

Similarly, we have

$$V_{BC} = V_{BN} - V_{CN} = V_{bn} - V_{cn} = V_{bc} = \sqrt{3}\, V_{ps} \angle^{\phi-90^o}$$

$$V_{CA} = V_{CN} - V_{AN} = V_{cn} - V_{an} = V_{ca} = \sqrt{3}\, V_{ps} \angle^{\phi+150^o}$$

Note that the magnitudes of these three line voltages are of the same value. The magnitude of each Y-load line voltage is denoted as V_{LY}. We have

$$V_{LY} = \sqrt{3} \cdot V_{pY} \qquad \text{7.2-2}$$

Combining **Equations 7.2-1** and **7.2-2**, we have

$$V_{LY} I_{LY} = \sqrt{3} \cdot V_{pY} I_{pY} \qquad \text{7.2-3}$$

Example 7.1. For a Y-Y balanced power system, the phase voltage of the source is given as 120 V. The impedance in each transmission line is given as $1+j$ Ω and the load impedance is given as $20+j10$ Ω. Find the line current and the line voltage of the load.

Solution: Due to the balance nature of the 3-phase power system, the calculation can be done using one phase as shown in **Figure 7.6**.

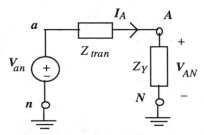

Figure 7.6. One phase of a 3-phase Y-Y balanced power system

The line current is found as follows:

$$I_A = \frac{V_{an}}{Z_{tran} + Z_Y} = \frac{120\angle^\phi}{(1+j)+(20+j10)}$$

$$= \frac{120\angle^\phi}{21+j11} = \frac{120\angle^\phi}{23.7\angle^{27.65^\circ}} = 5.06\angle^{\phi-27.65^\circ} \ A$$

Hence,

$$I_{LY} = I_{pY} = 5.06 \ A$$

The phase voltage of the load is found to be

$$V_{AN} = I_A Z_Y = 5.06\angle^{\phi-27.65^o} (20 + j10)$$
$$= 5.06\angle^{\phi-27.65^o} \times 22.4\angle^{26.57^o} = 113\angle^{\phi-1.08^o} V$$

Therefore,

$$V_{LY} = \sqrt{3} V_{pY} = \sqrt{3} \times 113 = 196 V$$

7.2.2 Balanced Δ-Loads

When a balance Δ-load is connected to a 3-phase source through transmission lines as shown in **Figure 7.7**, we have what is known as a Y-Δ balanced power system.

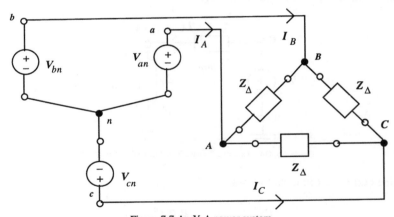

Figure 7.7 An Y-Δ power system

For a 3-phase Δ-load, Voltage V_{AB} is the voltage across the impedance $Z_Δ$ connected between Nodes A and B. This voltage is known as a phase voltage of the Δ-load. If the impedance of each transmission line is zero, we have

$$V_{AB} = V_{ab} = V_{Ls}\angle^\delta$$

Similarly, we have

$$V_{BC} = V_{bc} = V_{Ls}\angle^{\delta-120^o}$$

$$V_{CA} = V_{ca} = V_{Ls} \angle^{\delta+120^o}$$

Note that the magnitudes of these phase voltages are of the same value. The magnitude of each Δ-load phase voltage is denoted as $V_{p\Delta}$. Since each phase voltage is also the voltage across two of the three transmission lines, it is also known as a line voltage of the Δ-load. The magnitude of a Δ-load line voltage is denoted as $V_{L\Delta}$. Hence, we have

$$V_{p\Delta} = V_{L\Delta} \qquad\qquad 7.2\text{-}4$$

The current flowing through the impedance between Nodes A & B can be found as follows:

$$I_{AB} = \frac{V_{AB}}{Z_\Delta} = \frac{V_{p\Delta}\angle^{\delta}}{\rho_\Delta \angle^{\theta_\Delta}} = \frac{V_{p\Delta}}{\rho_\Delta}\angle^{\delta-\theta_\Delta} = I_{p\Delta}\angle^{\delta-\theta_\Delta}$$

Similarly, the currents flowing through the impedance between Nodes B & C and between Nodes C & A can be found as follows:

$$I_{BC} = \frac{V_{BC}}{Z_\Delta} = \frac{V_{p\Delta}\angle^{\delta-120^o}}{\rho_\Delta \angle^{\theta_\Delta}} = \frac{V_{p\Delta}}{\rho_\Delta}\angle^{\delta-\theta_\Delta-120^o} = I_{p\Delta}\angle^{\delta-\theta_\Delta-120^o}$$

$$I_{CA} = \frac{V_{CA}}{Z_\Delta} = \frac{V_{p\Delta}\angle^{\delta+120^o}}{\rho_\Delta \angle^{\theta_\Delta}} = \frac{V_{p\Delta}}{\rho_\Delta}\angle^{\delta-\theta_\Delta+120^o} = I_{p\Delta}\angle^{\delta-\theta_\Delta+120^o}$$

These currents are known as the phase currents of the Δ-load, and they are of the same magnitude. The magnitude of each Δ-load phase current is denoted as $I_{p\Delta}$.

Current I_A is the current flowing in one of the transmission lines. It can be found as follow:

$$I_A = I_{AB} - I_{CA} = I_{p\Delta}\angle^{\delta-\theta_\Delta} - I_{p\Delta}\angle^{\delta-\theta_\Delta+120^o}$$

$$= I_{p\Delta}\angle^{\delta-\theta_\Delta} \times [1 - 1\angle^{120^o}] = I_{p\Delta}\angle^{\delta-\theta_\Delta} \times [1-(-\frac{1}{2}+j\frac{\sqrt{3}}{2})]$$

$$= I_{p\Delta}\angle^{\delta-\theta_\Delta} \times \sqrt{3}\angle^{-30^o} = \sqrt{3}I_{p\Delta}\angle^{\delta-\theta_\Delta-30^o} = I_{L\Delta}\angle^{\delta-\theta_\Delta-30^o}$$

Similarly, we have

$$I_B = I_{BC} - I_{AB} = I_{p\Delta} \angle^{\delta-\theta_\Delta-120^o} - I_{p\Delta} \angle^{\delta-\theta_\Delta} = I_{L\Delta} \angle^{\delta-\theta_\Delta-150^o}$$

$$I_C = I_{CA} - I_{BC} = I_{p\Delta} \angle^{\delta-\theta_\Delta+120^o} - I_{p\Delta} \angle^{\delta-\theta_\Delta-120^o} = I_{L\Delta} \angle^{\delta-\theta_\Delta+90^o}$$

Note that

$$I_{L\Delta} = \sqrt{3} \cdot I_{p\Delta} \qquad\qquad 7.2\text{-}5$$

Applying Kirchhoff's current law to the supernode formed by the entire Δ-load, we have

$$I_A + I_B + I_C = 0$$

Combining **Equations 7.2-4 and 7.2-5**, we have

$$V_{L\Delta} I_{L\Delta} = V_{p\Delta} \sqrt{3} \cdot I_{p\Delta} = \sqrt{3} \cdot V_{p\Delta} I_{p\Delta} \qquad\qquad 7.2\text{-}6$$

Example 7.2. For a Y-Δ balanced power system, the phase voltage of the source is given as 440 V. The impedance in each transmission line is negligible and the load impedance is given as 12+j8 Ω. Find the line current of the Δ-load.

Solution: Since the impedance in each transmission line is negligible, the line voltage of the Δ-load is the line voltage of the source. Hence,

$$V_{L\Delta} = V_{Ls} = \sqrt{3} \cdot V_{ps} = \sqrt{3} \times 440 = 762\ V$$

The line current is found as follows:

$$I_{L\Delta} = \sqrt{3} \cdot I_{p\Delta} = \sqrt{3} \cdot \frac{V_{p\Delta}}{\rho_\Delta} = \sqrt{3} \cdot \frac{762}{\sqrt{12^2 + 8^2}} = 91.5\ A$$

It should be noted that the calculation is not as straightforward if the impedance in a transmission line is not negligible. This subject is deferred and is discussed in the following section.

7.2.3 Balanced Three-Phase Load Transformation

A balanced Δ-load is equivalent to a balanced Y-load as shown in **Figure 7.8** if we have

$$Z_Y = \frac{Z_\Delta}{3} \qquad\qquad 7.2\text{-}7$$

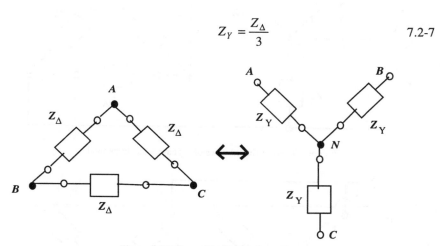

Figure 7.8 Balanced loads transformation

Such a transformation when properly applied can simplify calculation as in the following example.

Example 7.3. A 3-phase source is supplying power to two balanced loads, an Y-load and a Δ-load, as shown in **Figure 7.9**. The line voltage of the source is given as 440 V. The impedance in each transmission line is negligible, the load impedance of the Y-load is given as $5+j2$ Ω, and the load impedance of the Δ-load is given as $18+j9$ Ω. Find the line current of the 3-phase source.

Solution: To simplify the calculation, the Δ-load is first transformed into an Y-load as shown in **Figure 7.10**.

Since the impedance in each transmission line is negligible, the line voltage of the source is the line voltage of the loads in parallel. Without losing generality, we can assume that

$$V_{AN} = V_{an} = \frac{440}{\sqrt{3}} \angle 0^o \; V$$

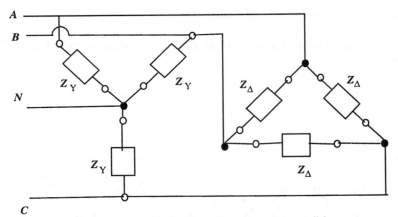

Figure 7.9 Two different types of 3-phase loads in parallel

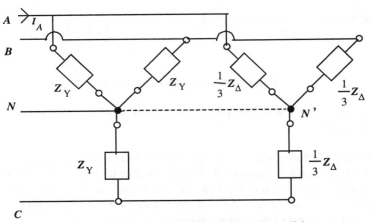

Figure 7.10 Two 3-phase balanced Y-loads in parallel

Due to the balanced nature of Y-load, the voltage between Nodes A and N is equal to the voltage between Nodes A and N'. The line current of the source can be found as follows:

$$I_A = \frac{V_{AN}}{Z_Y} + \frac{V_{AN'}}{Z_\Delta/3} = \frac{440}{\sqrt{3}} \angle 0^\circ \left[\frac{1}{5+j2} + \frac{1}{(18+j6)/3} \right]$$

$$= \frac{440}{\sqrt{3}} \left(\frac{5-j2}{29} + \frac{6-j2}{40} \right) = \frac{440}{\sqrt{3}} (0.322 - j0.119)$$

$$= \frac{440}{\sqrt{3}} \times 0.343 \angle^{-20.28^o} = 87.1 \angle^{-20.28^o} \ A$$

The line current of the source is thus found to be

$$I_{Ls} = 87.1 A$$

7.2.4 Three-Phase Power Calculation

The power absorbed by a 3-phase balanced load can be found by studying the power absorption of a single phase first. The power absorption of a single phase multiplied by three gives rise to the total power absorbed by the 3-phase load. The derivation for an Y-load is carried out first in the following paragraph. Similar results can be deduced for Δ-load.

The power absorbed by a single phase of an Y-load is given as follows:

$$P_{pY} = V_{pY} I_{pY} \cos\theta_Y$$

Hence, the power absorbed by the Y-load is found to be

$$P_{3pY} = 3P_{pY} = 3V_{pY} I_{pY} \cos\theta_Y = \sqrt{3} \cdot V_{LY} I_{LY} \cos\theta_Y$$

Similarly, the reactive power absorbed by the Y-load is given by

$$Q_{3pY} = 3Q_{pY} = 3V_{pY} I_{pY} \sin\theta_Y = \sqrt{3} \cdot V_{LY} I_{LY} \sin\theta_Y$$

The complex power absorbed by the Y-load is thus equal to

$$S_{3pY} = P_{3pY} + jQ_{3pY} = 3P_{pY} + j3Q_{pY} = 3(P_{pY} + jQ_{pY})$$
$$= 3I_{pY}^2 Z_Y = 3I_{LY}^2 Z_Y$$

The power absorbed by a single phase of a Δ-load is given as follows:

$$P_{p\Delta} = V_{p\Delta} I_{p\Delta} \cos\theta_\Delta$$

Hence, the power absorbed by the Δ-load is found to be

$$P_{3p\Delta} = 3P_{p\Delta} = 3V_{p\Delta}I_{p\Delta}\cos\theta_\Delta = \sqrt{3}\cdot V_{L\Delta}I_{L\Delta}\cos\theta_\Delta$$

Similarly, the reactive power absorbed by the Δ-load is given by

$$Q_{3p\Delta} = 3Q_{p\Delta} = 3V_{p\Delta}I_{p\Delta}\sin\theta_\Delta = \sqrt{3}\cdot V_{L\Delta}I_{L\Delta}\sin\theta_\Delta$$

The complex power absorbed by the Δ-load is thus equal to

$$S_{3p\Delta} = P_{3p\Delta} + jQ_{3p\Delta} = 3P_{p\Delta} + j3Q_{p\Delta} = 3(P_{p\Delta} + jQ_{p\Delta})$$
$$= 3I_{p\Delta}{}^2Z_\Delta = I_{L\Delta}{}^2Z_\Delta$$

Example 7.4. For a Y-Y balanced system, the line voltage of the source is 208 V. There is zero impedance in each of the transmission lines. The power absorbed by the load is 15.5 *kW* at power factor of 0.83, lagging. Determine the impedance of a phase of the Y-load.

Solution: Since we have

$$P_{3pY} = \sqrt{3}\cdot V_{LY}I_{LY}\cos\theta_Y,$$

the line current of the load is found to be

$$I_{LY} = \frac{P_{3pY}}{\sqrt{3}\cdot V_{LY}\cos\theta_Y} = \frac{15500}{\sqrt{3}\times208\times.0.83} = 51.8\,A$$

For the Y-load, the phase current is equal to the line current. The impedance of a phase of the Y-load is found to be

$$Z_Y = \frac{V_{pY}}{I_{pY}}\angle(+\cos^{-1}0.83) = \frac{208/\sqrt{3}}{51.8}\angle^{33.90^o} = 2.32\angle^{33.90^o}\,\Omega$$

Alternative Solution: The reactive power of the Y-load is found to be

$$Q_{3pY} = \sqrt{3}\cdot V_{LY}I_{LY}\sin\theta_Y = \frac{P_{3pY}}{\cos\theta_Y}\sin\theta_Y = P_{3pY}\tan\theta_Y$$
$$= 15500\times\tan(+\cos^{-1}0.83) = 10416\,VAR$$

Since the complex power of the Y-load is given as

$$S_{3pY} = P_{3pY} + jQ_{3pY} = 3I_{LY}^2 Z_Y,$$

the impedance of a phase of the Y-load is found to be

$$Z_Y = \frac{S_{3pY}}{3I_{LY}^2} = \frac{15500 + j10416}{3 \times 51.8^2} = 1.926 + j1.294 = 2.320\angle 33.90^o \ \Omega$$

Example 7.5. For a Y-Δ balanced system, the line voltage of the Δ-load is specified to be 440 V. The impedance in each transmission line is found to be $1 + j0.5 \ \Omega$. The impedance of a phase of the Δ-load is found to be $30 + j21 \ \Omega$. Find the total power supplied by the source and the line voltage of the source.

Solution: A phase current of the Δ-load is found as follows:

$$I_{AB} = \frac{V_{AB}}{Z_\Delta} = \frac{440\angle 0^o}{30 + j21} = 12.02\angle -34.99^o \ A$$

The line current of the Δ-load is therefore equal to

$$I_{L\Delta} = \sqrt{3} \cdot I_{p\Delta} = \sqrt{3} \times 12.02 = 20.82 \ A$$

The total real power supplied by the source is found to be

$$P_{3pS} = 3I_{L\Delta}^2 \times 1 + 3I_{p\Delta}^2 \times 10 = 3 \times 20.82^2 \times 1 + 3 \times 12.02^2 \times 30 = 14.30\,W$$

The total reactive power supplied by the source is found to be

$$Q_{3pS} = 3I_{L\Delta}^2 \times 0.5 + 3I_{p\Delta}^2 \times 7 = 3 \times 20.82^2 \times 0.5 + 3 \times 12.02^2 \times 21 = 9.75 \times 10^3 \ VAR$$

The line voltage of the source is thus found to be

$$V_{LS} = \frac{|S_{3pS}|}{\sqrt{3} \cdot I_{LS}} = \frac{|14.30 + j9.75| \times 10^3}{\sqrt{3} \times 20.82} = 480\,V$$

Example 7.6. A balanced 3-phase source is supplying power at a line voltage of 2200 V, *rms*, to the following three loads:

Load 1:   36 *kVA* at 0.8 *pf*, lagging

Load 2:   18 *kW* at 0.6 *pf*, leading

Load 3:   15 *kW* and 14 *kVA*

Find the line current of the 3-phase source and the power factor of the combined loads.

Solution: The total complex power of the combined loads is found to be

$$S_{3p} = 36\angle^{+\cos^{-1}0.8} + [18 + j18 \times \tan(-\cos^{-1}0.6)] + [15 + j14]$$
$$= 61.8 + j11.6\, kVA$$

The power factor of the combined loads is therefore

$$pf = \cos(\tan^{-1}\frac{11.6}{61.8}) = 0.983$$

The line current of the source is found to be

$$I_{LS} = \frac{|S_{3p}|}{\sqrt{3}V_{LS}} = \frac{|61.8 + j11.6| \times 10^3}{\sqrt{3} \times 2200} = 16.5\, A$$

Problems

7.2-1. Solve the following problems:
(a) A 3-phase balanced Y-source is delivering real power of 78 *kW* to a load with a leading power factor of 0.67 and at a line current of 90 *A*. Determine the line voltage of the Y-source.
(b) The line current of a 3-phase balanced Y-load is at 130 *A*. The impedance per phase $Z_Y = 5.6 + j4.3$ Ω. Determine the line voltage of the Y-load.
(c) The line voltage of a 3-phase balanced Δ-load is at 440 *V*. The impedance per phase $Z_\Delta = 10 + j7.5$ Ω. Determine the line current of the Δ-load.

7.2-2. A balanced Δ-connected load having an impedance of $135 + j72$ Ω per phase is fed from a line having an impedance of $3 + j5$ Ω per phase. The line-to-line voltage at the sending end of the line is 1100 V, rms. Calculate the line current and the line voltage of the Δ-load.

7.2-3. In a balanced Y-Δ three-phase system, the source line voltage is 330 V, rms. The impedance per phase of the delta load is $30 + j27$ Ω. The line impedance is $2 + j3$ Ω. Find the phase current of the delta load.

7.2-4. A balanced 3-phase distribution line is used to supply four balanced loads that are connected in parallel. The four loads are

$L_1 = 13$ kVA at 0.85 pf, lagging $L_2 = 25$ kVA at 0.93 pf, leading
$L_3 = 12$ kW and 7 $kVAR$ $L_4 = 16$ kW at unity pf

The line voltage at the loads is 440 V. Find the line current in the distribution line. What is the impedance per phase of an equivalent Δ-load?

7.2-5. The wire impedance per phase in each transmission line of a three-phase system is $Z_w = 5 + j3$ Ω. The lines feed three balanced loads that are connected in parallel. The loads are as follows:

Load 1: 37 kW, at a 0.72 lagging power factor

Load 2: 64 kVA, at a 0.83 leading power factor

Load 3: 55 kW and 29 $kVAR$

The line voltage at the load end is 3300 V, rms. Find the line current and the line voltage at the source end.

7.2-6. In a balanced Y-Δ three-phase system, the source line voltage is 330 V, rms. The impedance per phase of the delta load is $30 + j27$ Ω. The line impedance is $2 + j3$ Ω. Find the phase current of the Δ-load.

7.2-7. A balanced three-phase source serves the following loads:

Load 1: 200 kVA at 0.88 pf lagging
Load 2: 170 kVA at 0.76 pf leading
Load 3: 150 kW at unity pf

The load line voltage is 2200 V, rms. The line impedance is $0.1 + j0.2$ Ω. Find the line voltage and power factor at the source.

Chapter 8
Two-Port Networks

Many devices are equipped with more than a pair of terminals. These devices can not be properly modeled as one-port networks. The devices that are a level more complicated than one-ports are the so-called two-port networks. A two-port network contains two pair of terminals for connection to other circuit devices. Transformers and transistors are among many circuit devices that can be modeled by two-port networks.

8.1 Transformers

Transformers are widely utilized in the power industry for power transmission, in personal computers for power supplies, and in amplifiers for impedance matching. The operating mechanism of a transformer is an ever-changing magnetic field.

8.1.1 Ideal Transformer

In **Figure 8.1**, there are two sets of wires wrapped around a core. They are referred to as coils or windings. The core is generally made of iron. When there is current flowing in either wire, a magnetic field is formed inside and outside the core. With an iron core, the strength of the magnetic field inside is much stronger than the strength of the magnetic field outside. For simplicity, it is assumed that the magnetic field is confined within the core and denoted with the magnetic flux ϕ. When the magnetic field is changing, there are voltages induced in both wires as shown. The voltages induced are opposing the changing of the magnetic field.

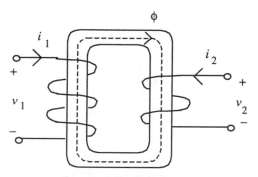

Figure 8.1 The basic configuration of a transformer

The voltages induced are proportional to the rate of the changing flux as follows:

$$v_1 = n_1 \frac{d\phi}{dt}$$

$$v_2 = n_2 \frac{d\phi}{dt}$$

where n_1 and n_2 are numbers of turns in Coil 1 and Coil 2, respectively. Taking the ratio of the two equations above, we thus have the equation that governs the voltage characteristic of an ideal transformer, namely,

$$\frac{v_1}{v_2} = \frac{n_1}{n_2} \qquad\qquad 8.1\text{-}1$$

The power absorbed by a transformer is given as follows:

$$p_{in} = v_1 i_1 + v_2 i_2$$

This power usually turns into heat and becomes useless and is hence referred to as power loss. When a transformer is operating at full load, the amount of power loss in the transformer is negligible compared to the power it transmits. Under such a condition, it is reasonable to assume that there is no power loss in a transformer, particularly in an ideal one. If the power loss is assumed zero, we then have

$$\frac{i_2}{i_1} = -\frac{v_1}{v_2} = -\frac{n_1}{n_2} \qquad\qquad 8.1\text{-}2$$

An ideal transformer with the characteristics governed by **Equations 8.1-1** and **8.1-2** is identified by the circuit symbol as shown in **Figure 8.2**.

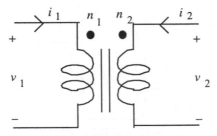

Figure 8.2. The symbol of an ideal transformer

The two dots in the symbol indicate how voltages are induced in the transformer. When an increasing current enters a coil of the transformer at a dotted terminal, it induces a voltage in the second coil with a polarity positive at the dotted terminal of the second coil. This notation is known as the dot convention.

8.1.2 Transformer Application

Since voltages in the transformer coils are induced by a changing magnetic field, transformers are not desirable in DC circuits. When sources in a circuit containing a transformer are sinusoidal and of the same frequency, the transformer can be modeled with the circuit shown in **Figure 8.3**. In the figure, voltage and current signals are represented by phasors.

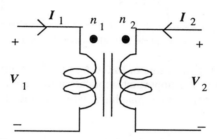

Figure 8.3. The AC model of an ideal transformer

Equations 8.1-1 and **8.1-2** then become as follows:

$$\begin{cases} \dfrac{V_1}{V_2} = \dfrac{n_1}{n_2} \\[4mm] \dfrac{I_2}{I_1} = -\dfrac{n_1}{n_2} \end{cases} \qquad 8.1\text{-}3$$

A common AC circuit containing a transformer is shown in **Figure 8.4**. In the circuit shown, a load represented by the impedance Z_L is connected to Coil 2 on the right. This coil is normally referred to as the secondary winding. Coil 1 on the left is called the primary winding which is normally connected to a source.

The equivalent impedance seen by the source in **Figure 8.4** are found to be

$$Z_{eq} = \frac{V_1}{I_1} = \frac{\frac{n_1}{n_2}V_2}{-\frac{n_2}{n_1}I_2} = \left(\frac{n_1}{n_2}\right)^2 \frac{V_2}{-I_2} = \left(\frac{n_1}{n_2}\right)^2 Z_L$$

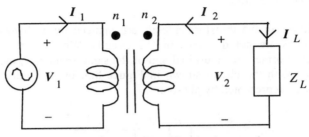

Figure 8.4. A typical circuit configuration for a transformer

By properly adjusting the value of the turn ratio, the equivalent impedance can be modified accordingly. An important application of such a circuit is found in a situation known as the impedance matching, which is illustrated in the following example.

Example 8.1 Determine the turn ratio of the transformer in **Figure 8.5** so that the maximum power can be transferred to the load represented by the resistor R_L.

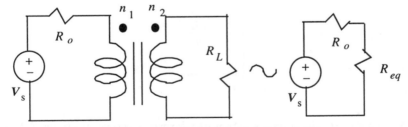

Figure 8.5. A transformer circuit and its equivalent circuit

Solution: Since an ideal transformer does not consume power, whatever power enters the primary winding is transmitted to the load through the secondary winding. Consequently, the power absorbed by the equivalent resistor R_{eq} is equal to the power absorbed by the load resistor R_L. According to Thevenin Theorem, the maximum power is transferred to the equivalent resistor when the output resistance is equal to the equivalent resistance, i.e.,

$$R_o = R_{eq} = \left(\frac{n_1}{n_2}\right)^2 R_L$$

Hence, for maximum power transfer, the turn ratio must be as follows:

$$\frac{n_1}{n_2} = \sqrt{\frac{R_o}{R_L}}$$

For a typical value of 5kΩ for the output resistance of a transistor amplifier and 8Ω for the load resistance of a speaker, the turn ratio should be 25.

Example 8.2 Find the voltage V_L in the circuit of **Figure 8.6**.

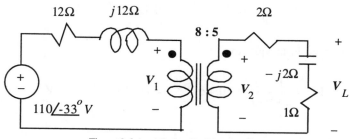

Figure 8.6 An AC circuit with a transformer

Solution: Firstly, let's find the voltage V_1. For V_1, we use the equivalent circuit in **Figure 8.7**.

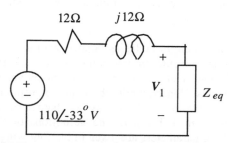

Figure 8.7 The equivalent circuit for the primary winding

The equivalent impedance as seen from the primary winding is found to be

$$Z_{eq} = \left(\frac{n_1}{n_2}\right)^2 (2 - j2 + 1) = \left(\frac{8}{5}\right)^2 (2 - j2 + 1) = 7.68 - j5.12 \quad \Omega$$

The voltage V_1 is thus equal to

$$V_1 = \frac{7.68 - j5.12}{12 + j12 + 7.68 - j5.12} \times 110\angle^{-33^o}$$

$$= \frac{9.23\angle^{-33.69^o} \times 110\angle^{-33^o}}{20.85\angle^{19.27^o}} = 48.7\angle^{-85.96^o} \quad V$$

The voltage V_2 is therefore equal to

$$V_2 = \frac{5}{8} \times 48.7\angle^{-85.96^o} = 30.4\angle^{-85.96^o} \quad V$$

Finally, the voltage V_L is equal to

$$V_L = \frac{1 - j2}{2 + 1 - j2} \times 30.4\angle^{-85.96^o} = 18.85\angle^{-115.50^o} \quad V$$

8.2 Two-Port Parameters

In the rest of this chapter, we assume that all the sources are sinusoidal and of the same frequency. Consequently, all the voltage and current signals are represented by phasors. Meanwhile, a passive circuit element is represented by the corresponding impedance.

There are four circuit variables associated with a two-port network containing no independent sources inside. In describing the characteristics of the two-port, two out of the four variables can be chosen as independent variables while the other two as dependent variables. Consequently, there are six different ways of describing the characteristics of a two-port. Each way is characterized by a set of two-port parameters as shown in **Table 8.1**. The definitions of these parameters and the methods to find these parameters are given in the following sections.

8.2.1 Impedance Parameters

The equations that define the impedance parameters (also denoted as z-parameters) of a two-port network are

$$V_1 = z_{11}I_1 + z_{12}I_2 \qquad\qquad 8.2\text{-}1$$

$$V_2 = z_{21}I_1 + z_{22}I_2 \qquad\qquad 8.2\text{-}2$$

Table 8.1 Various types of two-port parameters			
types	*indep. variables*	*dep. variables*	*equations*
impedance	I_1 and I_2	V_1 and V_2	$V_1=z_{11}I_1+z_{12}I_2$ $V_2=z_{21}I_1+z_{22}I_2$
admittance	V_1 and V_2	I_1 and I_2	$I_1=y_{11}V_1+y_{12}V_2$ $I_2=y_{21}V_1+y_{22}V_2$
hybrid	I_1 and V_2	V_1 and I_2	$V_1=h_{11}I_1+h_{12}V_2$ $I_2=h_{21}I_1+h_{22}V_2$
inverse hybrid	V_1 and I_2	I_1 and V_2	$I_1=g_{11}V_1+g_{12}I_2$ $V_2=g_{21}V_1+g_{22}I_2$
transmission	V_2 and I_2	V_1 and I_1	$V_1=AV_2-BI_2$ $I_1=CV_2-DI_2$
inverse transmission	V_1 and I_1	V_2 and I_2	$V_2=A'V_1-B'I_1$ $I_2=C'V_1-D'I_1$

Given a two-port network, the impedance parameters of the network can be found as follows:

When I_2 is set to zero by opening the circuit at the second port, **Equations 8.2-1** and **8.2-2** become

$$V_1 = z_{11}I_1$$

$$V_2 = z_{21}I_1$$

Hence,

$$z_{11} = \left.\frac{V_1}{I_1}\right|_{I_2=0} \qquad\qquad 8.2\text{-}3$$

$$z_{21} = \left.\frac{V_2}{I_1}\right|_{I_2=0} \qquad\qquad 8.2\text{-}4$$

The subscripts in **Equations 8.2-3** and **8.2-4**. emphasize the condition at which the parameter z_{11} and z_{21} are to be found. A way to utilize these equations is to connect a current source with a current of one ampere at the first port while keeping the second port open. Once the corresponding values are found for the voltages, z_{11} has the same value as V_1 and z_{21} has the same value as V_2.

In a similar fashion, we have

$$z_{12} = \left. \frac{V_1}{I_2} \right|_{I_1=0} \qquad \text{8.2-5}$$

$$z_{22} = \left. \frac{V_2}{I_2} \right|_{I_1=0} \qquad \text{8.2-6}$$

Parameters z_{12} and z_{22} can be found by connecting a current source of one ampere at the second port while opening the first port. Once the corresponding values are found for the voltages, z_{12} has the same value as V_1 and z_{22} has the same value as V_2.

Example 8.3 Determine the z-parameters of the 2-port network shown in **Figure 8.6**.

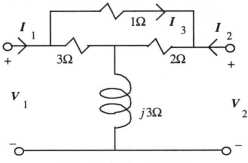

Figure 8.6. A 2-port network

Solution: For z_{11} and z_{21}, we open the circuit at the second port to set $I_2 = 0$. The parameter z_{11} is the same as the equivalent impedance at the first port. Hence, we have

$$z_{11} = \frac{3(1+2)}{3+(1+2)} + j3 = 1.5 + j3 \ \Omega$$

When $I_2 = 0$, the current in the 1Ω-resistor is found to be

$$I_3 = \frac{3}{3+(1+2)}I_1 = \frac{1}{2}I_1$$

and the voltage V_2 is found to be

$$V_2 = 2I_3 + j3I_1 = I_1 + j3I_1 = (1+j3)I_1$$

Hence, we have

$$z_{21} = \frac{V_2}{I_1}\bigg|_{I_2=0} = \frac{(1+j3)I_1}{I_1} = 1+j3 \ \Omega$$

For z_{12} and z_{22}, we open the circuit at the first port to set $I_1 = 0$. The parameter z_{22} is the same as the equivalent impedance at the second port. Hence, we have

$$z_{22} = \frac{2(1+3)}{2+(1+3)} + j3 = \frac{4}{3} + j3 \ \Omega$$

When $I_1 = 0$, the current in the 1Ω-resistor is found to be

$$I_3 = \frac{-2}{2+(1+3)}I_2 = \frac{-1}{3}I_1$$

and the voltage V_1 is found to be

$$V_1 = -3I_3 + j3I_2 = I_1 + j3I_1 = (1+j3)I_1$$

Hence, we have

$$z_{12} = \frac{V_1}{I_2}\bigg|_{I_1=0} = \frac{(1+j3)I_2}{I_2} = 1+j3 \ \Omega$$

Note that z_{12} and z_{21} have the same value. A 2-port network with this property is known as a reciprocal 2-port. Any passive 2-port containing no dependent source is a reciprocal 2-port.

8.2.2 Admittance Parameters

The equations that define the admittance parameters (also denoted as y-parameters) of a two-port network are

$$I_1 = y_{11}V_1 + y_{12}V_2 \qquad\qquad 8.2\text{-}7$$

$$I_2 = y_{21}V_1 + y_{22}V_2 \qquad\qquad 8.2\text{-}8$$

Given a two-port network, the admittance parameters of the network can be found as follows:

When V_2 is set to zero by shorting the two terminals together at the second port, **Equations 8.2-7** and **8.2-8** become

$$I_1 = y_{11}V_1$$

$$I_2 = y_{21}V_1$$

Hence,

$$y_{11} = \left.\frac{I_1}{V_1}\right|_{V_2=0} \qquad\qquad 8.2\text{-}9$$

$$y_{21} = \left.\frac{I_2}{V_1}\right|_{V_2=0} \qquad\qquad 8.2\text{-}10$$

The subscripts **in Equations 8.2-9** and **8.2-10** emphasize the condition at which the parameter y_{11} and y_{21} are to be found. A way to utilize these equations is to connect a voltage source with a voltage of one volt at the first port while shorting the two terminals together at the second port. Once the corresponding values are found for the currents, y_{11} has the same value as I_1 and y_{21} has the same value as I_2.

In a similar fashion, we have

$$y_{12} = \left.\frac{I_1}{V_2}\right|_{V_1=0} \qquad\qquad 8.2\text{-}11$$

$$y_{22} = \frac{I_2}{V_2}\bigg|_{V_1=0} \qquad\qquad 8.2\text{-}12$$

Parameters y_{12} and y_{22} can be found by connecting a voltage source of one volt at the second port while shorting the two terminals together at the first port. Once the corresponding values are found for the currents, y_{12} has the same value as I_1 and y_{22} has the same value as I_2.

Example 8.4 Determine the *y*-parameters of the 2-port network shown in **Figure 8.6**.

Solution: For y_{11} and y_{21}, we short the two terminals together at the second port to set $V_2 = 0$. The 2-port network in **Figure 8.6** then becomes as the one shown in **Figure 8.7**, with node voltages assigned for nodal analysis.

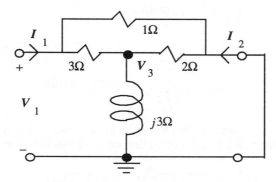

Figure 8.7. A circuit for finding a couple of y-parameters for the 2-port network

Applying KCL at Node **3**, we have

$$\frac{V_3 - V_1}{3} + \frac{V_3}{2} + \frac{V_3}{j3} = 0$$

Solving V_3 in terms of V_1, we have

$$V_3 = \frac{V_1}{2.5 - j} = (\frac{2.5}{7.25} + j\frac{1}{7.25})V_1$$

The current I_1 is then found to be

$$I_1 = \frac{V_1}{1} + \frac{V_1 - V_3}{3} = \left[1 + \frac{1}{3} - \frac{1}{3}(\frac{2.5}{7.25} + j\frac{1}{7.25})\right]V_1$$
$$= (1.218 - j0.046)V_1$$

Hence the parameter y_{11} is found to be

$$y_{11} = \frac{I_1}{V_1}\bigg|_{V_2=0} = \frac{(1.218 - j0.046)V_1}{V_1} = 1.218 - j0.046$$

The current I_2 is found to be

$$I_2 = -\frac{V_1}{1} - \frac{V_3}{2} = \left[-1 - \frac{1}{2}(\frac{2.5}{7.25} + j\frac{1}{7.25})\right]V_1$$
$$= (-1.172 - j0.069)V_1$$

Hence, we have

$$y_{21} = \frac{I_2}{V_1}\bigg|_{V_2=0} = -1.172 - j0.069$$

For y_{12} and y_{22}, we short the two terminals at the first port to set $V_1 = 0$. The 2-port network in **Figure 8.6** then becomes the one as shown in **Figure 8.8**.

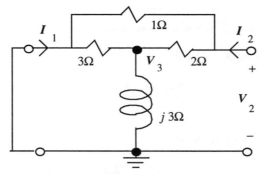

Figure 8.8. A circuit for finding a couple of y-parameters for the 2-port network

Applying KCL at Node **3**, we have

$$\frac{V_3 - V_2}{2} + \frac{V_3}{3} + \frac{V_3}{j3} = 0$$

Solving V_3 in terms of V_2, we have

$$V_3 = \frac{3V_1}{5 - j2} = (\frac{15}{29} + j\frac{6}{29})V_1$$

The current I_2 is then found to be

$$I_2 = \frac{V_2}{1} + \frac{V_2 - V_3}{2} = \left[1 + \frac{1}{2} - \frac{1}{2}(\frac{15}{29} + j\frac{6}{29})\right]V_2$$
$$= (1.241 - j0.103)V_2$$

Hence the parameter y_{22} is found to be

$$y_{22} = \frac{I_2}{V_2}\bigg|_{V_1=0} = \frac{(1.241 - j0.103)V_2}{V_2} = 1.241 - j0.103$$

The current I_1 is found to be

$$I_1 = -\frac{V_2}{1} - \frac{V_3}{3} = \left[-1 - \frac{1}{3}(\frac{15}{29} + j\frac{6}{29})\right]V_2$$
$$= (-1.172 - j0.069)V_2$$

Hence, we have

$$y_{12} = \frac{I_1}{V_2}\bigg|_{V_1=0} = \frac{(-1.172 - j0.069)V_2}{V_2} = -1.172 - j0.069$$

Note that y_{12} and y_{21} have the same value. This is true for all reciprocal 2-ports.

8.2.3 Hybrid Parameters

The equations that define the hybrid parameters (also denoted as h-parameters) of a two-port network are

$$V_1 = h_{11}I_1 + h_{12}V_2 \qquad\qquad 8.2\text{-}13$$

$$I_2 = h_{21}I_1 + h_{22}V_2 \qquad\qquad \text{8.2-14}$$

Given a two-port network, the hybrid parameters of the network can be found as follows:

When V_2 is set to zero by shorting the two terminals together at the second port, **Equations 8.2-13** and **8.2-14** become

$$V_1 = h_{11}I_1$$

$$I_2 = h_{21}I_1$$

Hence,

$$h_{11} = \frac{V_1}{I_1}\bigg|_{V_2=0} \qquad\qquad \text{8.2-15}$$

$$h_{21} = \frac{I_2}{I_1}\bigg|_{V_2=0} \qquad\qquad \text{8.2-16}$$

The subscripts in **Equations 8.2-15** and **8.2-16** emphasize the condition at which the parameter h_{11} and h_{21} are to be found. A way to utilize these equations is to connect a current source with a current of one ampere at the first port while shorting the two terminals together at the second port. Once the corresponding values are found for other circuit variables, h_{11} has the same value as V_1 and h_{21} has the same value as I_2.

In a similar fashion, we have

$$h_{12} = \frac{V_1}{V_2}\bigg|_{I_1=0} \qquad\qquad \text{8.2-17}$$

$$h_{22} = \frac{I_2}{V_2}\bigg|_{I_1=0} \qquad\qquad \text{8.2-18}$$

Parameters h_{12} and h_{22} can be found by connecting a voltage source of one volt at the second port while opening the the first port. Once the corresponding values are found for the other circuit variables, h_{12} has the same value as V_1 and h_{22} has the same value as I_2.

Unlike the z- or the y-parameters, h-parameters have mixed units. The first port of a 2-port network is conventionally called the input port and the second port is conventionally called the output port. The parameter h_{11} is the ratio of the input voltage and the input current and is with the units of impedance. It is known as the input impedance. The parameter h_{21} is the ratio of the output current and the input current and is hence without units. This parameter is known as the forward current ratio. The parameter h_{12} is the ratio of input voltage and the output voltage and hence it is also without units. This parameter is known as the reverse voltage ratio. The parameter h_{22} is the ratio of the output current and the output voltage and hence has the units of admittance. It is known as the output admittance.

Example 8.5 Determine the h-parameters of the 2-port network shown in **Figure 8.6**.

Solution: For h_{11} and h_{21}, we short the two terminals together at the second port to set $V_2 = 0$. The 2-port network in **Figure 8.6** then becomes the same network as shown in **Figure 8.7**. Hence, we have the exact equations as in the first part of the solution in **Example 8.4**.

$$\frac{V_3 - V_1}{3} + \frac{V_3}{2} + \frac{V_3}{j3} = 0$$

$$V_3 = \frac{V_1}{2.5 - j} = (\frac{2.5}{7.25} + j\frac{1}{7.25})V_1$$

$$I_1 = \frac{V_1}{1} + \frac{V_1 - V_3}{3} = \left[1 + \frac{1}{3} - \frac{1}{3}(\frac{2.5}{7.25} + j\frac{1}{7.25})\right]V_1$$
$$= (1.218 - j0.046)V_1$$

$$I_2 = -\frac{V_1}{1} - \frac{V_3}{2} = \left[-1 - \frac{1}{2}(\frac{2.5}{7.25} + j\frac{1}{7.25})\right]V_1$$
$$= (-1.172 - j0.069)V_1$$

The parameter h_{11} is found to be

$$h_{11} = \frac{V_1}{I_1}\bigg|_{V_2=0} = \frac{1}{\dfrac{I_1}{V_1}\bigg|_{V_2=0}} = \frac{1}{y_{11}} = \frac{1}{1.218 - j0.046}$$

Note that

$$h_{11} = \frac{V_1}{I_1}\bigg|_{V_2=0} \neq z_{11} = \frac{V_1}{I_1}\bigg|_{I_2=0}$$

The parameter h_{21} is found to be

$$h_{21} = \frac{I_2}{I_1}\bigg|_{V_2=0} = \frac{\dfrac{I_2}{V_1}\bigg|_{V_2=0}}{\dfrac{I_1}{V_1}\bigg|_{V_2=0}} = \frac{y_{21}}{y_{11}} = \frac{-1.172 - j0.069}{1.218 - j0.046}$$

For h_{12} and h_{22}, we open the circuit at the first port to set $I_1 = 0$. This is the same situation as in **Example 8.3** when z_{12} and z_{22} were found. Hence, we have

$$h_{12} = \frac{V_1}{V_2}\bigg|_{I_1=0} = \frac{\dfrac{V_1}{I_2}\bigg|_{I_1=0}}{\dfrac{V_2}{I_2}\bigg|_{I_1=0}} = \frac{z_{12}}{z_{22}} = \frac{1+j3}{\dfrac{4}{3}+j3}$$

$$h_{22} = \frac{I_2}{V_2}\bigg|_{I_1=0} = \frac{1}{\dfrac{V_2}{I_2}\bigg|_{I_1=0}} = \frac{1}{z_{22}} = \frac{1}{\dfrac{4}{3}+j3}$$

8.2.4 Inverse Hybrid Parameters

The equations that define the inverse hybrid parameters (also known as g-parameters) of a two-port network are

$$I_1 = g_{11}V_1 + g_{12}I_2 \qquad\qquad 8.2\text{-}19$$

$$V_2 = g_{21}V_1 + g_{22}I_2 \qquad \text{8.2-20}$$

Given a two-port network, the inverse hybrid parameters of the network can be found as follows:

When I_2 is set to zero by shorting the two terminals together at the second port, **Equations 8.2-19** and **8.2-20** become

$$I_1 = g_{11}V_1$$

$$V_2 = g_{21}V_1$$

Hence,

$$g_{11} = \left. \frac{I_1}{V_1} \right|_{I_2=0} \qquad \text{8.2-21}$$

$$g_{21} = \left. \frac{V_2}{V_1} \right|_{I_2=0} \qquad \text{8.2-22}$$

The subscripts in **Equations 8.2-21 and 8.2-22** emphasize the condition at which the parameters g_{11} and g_{21} are to be found. A way to utilize these equations is to connect a voltage source of one volt at the first port while keeping the circuit at the second port open. Once the corresponding values are found for other circuit variables, g_{11} has the same value as I_1 and g_{21} has the same value as V_2.

In a similar fashion, we have

$$g_{12} = \left. \frac{I_1}{I_2} \right|_{V_1=0} \qquad \text{8.2-23}$$

$$g_{22} = \left. \frac{V_2}{I_2} \right|_{V_1=0} \qquad \text{8.2-24}$$

Parameters g_{12} and g_{22} can be found by connecting a current source of one ampere at the second port while shorting the two terminals together at the first port. Once the corresponding values are found for other circuit variables, g_{12} has the same value as I_1 and g_{22} has the same value as V_2.

Similar to the h-parameters, the g-parameters have mixed units. The parameter g_{11} is the ratio of the input current and the input voltage and hence has the units of admittance. The parameter g_{21} is the ratio of the output voltage and the input voltage and hence is without units. The parameter g_{12} is the ratio of input current and the output current and is also without units. The parameter g_{22} is the ratio of the output voltage and the output current and hence has the units of impedance.

Example 8.6 Determine the g-parameters of the 2-port network shown in **Figure 8.6**.

Solution: For g_{11} and g_{21}, we open the circuit at the second port to set $I_2 = 0$. This is the same situation as in **Example 2** when z_{11} and z_{21} are found. Hence, we have

$$g_{11} = \frac{I_1}{V_1}\bigg|_{I_2=0} = \frac{1}{\dfrac{V_1}{I_1}\bigg|_{I_2=0}} = \frac{1}{z_{11}} = \frac{1}{1.5 + j3}$$

$$g_{21} = \frac{V_2}{V_1}\bigg|_{I_2=0} = \frac{\dfrac{V_2}{I_1}\bigg|_{I_2=0}}{\dfrac{V_1}{I_1}\bigg|_{I_2=0}} = \frac{z_{21}}{z_{11}} = \frac{1 + j3}{1.5 + j3}$$

For g_{12} and g_{22}, we short the two terminals together at the first port to set $V_1 = 0$. The 2-port network in **Figure 8.6** then becomes the same network as shown in **Figure 8.8**. Hence, we have the exact equations as in the second part of the solution in **Example 8.4**.

$$\frac{V_3 - V_2}{2} + \frac{V_3}{3} + \frac{V_3}{j3} = 0$$

$$V_3 = \frac{3V_1}{5 - j2} = (\frac{15}{29} + j\frac{6}{29})V_1$$

$$I_2 = \frac{V_2}{1} + \frac{V_2 - V_3}{2} = \left[1 + \frac{1}{2} - \frac{1}{2}(\frac{15}{29} + j\frac{6}{29})\right]V_2$$
$$= (1.241 - j0.103)V_2$$

$$I_1 = -\frac{V_2}{1} - \frac{V_3}{3} = \left[-1 - \frac{1}{3}(\frac{15}{29} + j\frac{6}{29})\right]V_2$$
$$= (-1.172 - j0.069)V_2$$

Hence, we have

$$g_{12} = \frac{I_1}{I_2}\bigg|_{V_1=0} = \frac{\dfrac{I_1}{V_2}\bigg|_{V_1=0}}{\dfrac{I_2}{V_2}\bigg|_{V_1=0}} = \frac{y_{12}}{y_{22}} = \frac{-1.172 - j0.069}{1.241 - j0.103}$$

$$g_{22} = \frac{V_2}{I_2}\bigg|_{V_1=0} = \frac{1}{\dfrac{I_2}{V_2}\bigg|_{V_1=0}} = \frac{1}{y_{22}} = \frac{1}{1.241 - j0.103}$$

8.2.5. Transmission Parameters

The equations that define the transmission parameters of a two-port network are

$$V_1 = AV_2 - BI_2 \qquad\qquad 8.2\text{-}25$$

$$I_1 = CV_2 - DI_2 \qquad\qquad 8.2\text{-}26$$

Given a two-port network, the transmission parameters of the network can be found as follows:

When I_2 is set to zero by opening the circuit at the second port, **Equations 8.2.25** and **8.2.26** become

$$V_1 = AV_2$$

$$I_1 = CV_2$$

Hence,

$$A = \frac{V_1}{V_2}\bigg|_{I_2=0} \qquad\qquad 8.2\text{-}27$$

$$C = \frac{I_1}{V_2}\bigg|_{I_2=0} \qquad\qquad 8.2\text{-}28$$

The subscripts in **Equations 8.2-27** and **8.2-28** emphasize the condition at which the parameter A and C are to be found. A way to utilize these equations is to connect a voltage source or a current source at the first port while keeping the circuit at the second port open.

In a similar fashion, we have

$$B = \frac{V_1}{-I_2}\bigg|_{V_2=0} \qquad\qquad 8.2\text{-}29$$

$$D = \frac{I_1}{-I_2}\bigg|_{V_2=0} \qquad\qquad 8.2\text{-}30$$

Parameters B and D can be found by connecting a current source or a voltage source at the first port while shorting the two terminals together at the second port.

Similar to the h-parameters and the g-parameters, transmission parameters are with mixed units. The parameter A is the ratio of the input voltage and the output voltage and is hence without units. The parameter C is the ratio of the input current and the output voltage and is hence with the units of admittance. The parameter B is the ratio of the input voltage and the negative of the output current and is hence with the units of impedance. The parameter D is the ratio of the input current and the negative of the output current and is hence without units.

Example 8.7 Determine the transmission parameters of the 2-port network shown in **Figure 8.6**.

Solution: For A and C, we open the circuit at the second port to set $I_2 = 0$. This is the same situation as in **Example 8.3** when z_{11} and z_{21} are found. Hence, we have

$$A = \frac{V_1}{V_2}\bigg|_{I_2=0} = \frac{\dfrac{V_1}{I_1}\bigg|_{I_2=0}}{\dfrac{V_2}{I_1}\bigg|_{I_2=0}} = \frac{z_{11}}{z_{21}} = \frac{1.5 + j3}{1 + j3}$$

$$C = \frac{I_1}{V_2}\bigg|_{I_2=0} = \frac{1}{\dfrac{V_2}{I_1}\bigg|_{I_2=0}} = \frac{1}{z_{21}} = \frac{1}{1+j3}$$

For B and D, we short the two terminals together at the second port to set $V_2 = 0$. The 2-port network in **Figure 8.6** then becomes as the very same one as shown in **Figure 8.7**. Once y_{11} and y_{21} are found, we have

$$B = \frac{V_1}{-I_2}\bigg|_{V_2=0} = -\frac{1}{\dfrac{I_2}{V_1}\bigg|_{V_2=0}} = -\frac{1}{y_{21}} = \frac{1}{1.172+j0.069}$$

$$D = \frac{I_1}{-I_2}\bigg|_{V_2=0} = -\frac{\dfrac{I_1}{V_1}\bigg|_{V_2=0}}{\dfrac{I_2}{V_1}\bigg|_{V_2=0}} = -\frac{y_{11}}{y_{21}} = \frac{1.218-j0.046}{1.172+j0.069}$$

8.2.6. Inverse Transmission Parameters

The equations that define the inverse transmission parameters of a two-port network are

$$V_2 = A'V_1 - B'I_1 \qquad\qquad 8.2\text{-}31$$

$$I_2 = C'V_1 - D'I_1 \qquad\qquad 8.2\text{-}32$$

Given a two-port network, the inverse transmission parameters of the network can be found as follows:

When I_1 is set to zero by opening the circuit at the first port, **Equations 8.2-31** and **8.2-32** become

$$V_2 = A'V_1$$

$$I_2 = C'V_1$$

Hence,

$$A' = \frac{V_2}{V_1}\bigg|_{I_1=0} \qquad\qquad 8.2\text{-}33$$

$$C' = \frac{I_2}{V_1}\bigg|_{I_1=0} \qquad\qquad 8.2\text{-}34$$

The subscripts in **Equations 8.2-33** and **8.2-34** emphasize the condition at which the parameter A' and C' are to be found. A way to utilize these equations is to connect a voltage source or a current source at the second port while keeping the circuit at the first port open.

In a similar fashion, we have

$$B' = \frac{V_2}{-I_1}\bigg|_{V_1=0} \qquad\qquad 8.2\text{-}35$$

$$D' = \frac{I_2}{-I_1}\bigg|_{V_1=0} \qquad\qquad 8.2\text{-}36$$

Parameters B' and D' can be found by connecting a current source or a voltage source at the second port while shorting the two terminals together at the first port.

Similar to the h-parameters and the g-parameters, inverse transmission parameters have mixed units. The parameter A' is the ratio of the output voltage and the input voltage and is without units. The parameter C' is the ratio of the output current and the input voltage and is with the units of admittance. The parameter B' is the ratio of the output voltage and the negative of the input current and is with the units of impedance. The parameter D' is the ratio of the output current and the negative of the input current and is without units.

Example 8.8 Determine the inverse transmission parameters of the 2-port network shown in **Figure 8.6**.

For A' and C', we open the circuit at the first port to set $I_1 = 0$. This is the same situation as in **Example 8.3** when z_{12} and z_{22} are found. Hence, we have

$$A' = \frac{V_2}{V_1}\bigg|_{I_1=0} = \frac{\dfrac{V_2}{I_2}\bigg|_{I_1=0}}{\dfrac{V_1}{I_2}\bigg|_{I_1=0}} = \frac{z_{22}}{z_{12}} = \frac{\dfrac{4}{3}+j3}{1+j3}$$

$$C' = \frac{I_2}{V_1}\bigg|_{I_1=0} = \frac{1}{\dfrac{V_1}{I_2}\bigg|_{I_1=0}} = \frac{1}{z_{12}} = \frac{1}{1+j3}$$

For B' and D', we short the two terminals together at the first port to set $V_1 = 0$. The 2-port network in **Figure 8.6** then becomes the same network as shown in **Figure 8.8**. Once y_{12} and y_{22} are found, we have

$$B' = \frac{V_2}{-I_1}\bigg|_{V_1=0} = -\frac{1}{\dfrac{I_1}{V_2}\bigg|_{V_1=0}} = -\frac{1}{y_{12}} = \frac{1}{1.172+j0.069}$$

$$D' = \frac{I_2}{-I_1}\bigg|_{V_1=0} = -\frac{\dfrac{I_2}{V_2}\bigg|_{V_1=0}}{\dfrac{I_1}{V_2}\bigg|_{V_1=0}} = -\frac{y_{22}}{y_{12}} = \frac{1.241-j0.103}{1.172+j0.069}$$

8.3 Parameter Conversions and Applications

When one set of parameters of a given 2-port network is found, other sets can be derived without further analyzing the network. The process of finding another set of parameters based on the knowledge of a given set is known as the parameter conversion.

8.3.1 Conversions

A set of parameters for a 2-port network and its defining equations can be utilized to derive another set of parameters. Although there are a total of 30 conversions, it is not necessary to study every one of them since every conversion can be derived in a similar fashion. The following example illustrates the methodology.

Example 8.9 Given the z-parameters, derive the transmission parameters.

Solution: From **Equations 8.2-1** and **8.2-2**, we have

$$z_{21}I_1 = V_2 - z_{22}I_2$$

If the parameter z_{21} is not zero, we have

$$I_1 = \frac{V_2 - z_{22}I_2}{z_{21}} = CV_2 - DI_2$$

Hence, we have

$$C = \frac{1}{z_{21}} \quad \text{and} \quad D = \frac{z_{22}}{z_{21}}$$

Furthermore,

$$V_1 = z_{11}I_1 + z_{12}I_2 = z_{11}\frac{V_2 - z_{22}I_2}{z_{21}} + z_{12}I_2$$

$$= \frac{z_{11}}{z_{21}}V_2 - \frac{z_{11}z_{22} - z_{21}z_{12}}{z_{21}}I_2 = AV_2 - BI_2$$

Therefore,

$$A = \frac{z_{11}}{z_{21}} \quad \text{and} \quad B = \frac{z_{11}z_{22} - z_{21}z_{12}}{z_{21}}$$

8.3.2 Application

Two-port network models are widely used in electronics to represent devices such as bipolar junction transistors and field effect transistors. The configuration shown in **Figure 8.9** is typically utilized in an amplifier circuit containing a transistor. The two-port network is identified by the set of h-parameters shown in the figure.

The amplification of the amplifier is normally defined as the ratio of V_2 and V_1 and is derived in the following. In addition to **Equations 8.2-13** and **8.2-14**, we have

$$V_2 = -R_L I_2$$

Together with **Equation 8.2-14**, we have

$$I_2 = -\frac{V_2}{R_L} = h_{21}I_1 + h_{22}V_2$$

Hence, we have

$$-\frac{V_2}{R_L} - h_{22}V_2 = h_{21}I_1$$

i.e.,

$$I_1 = -\frac{\dfrac{1}{R_L} + h_{22}}{h_{21}}V_2 = -\frac{1 + R_L h_{22}}{R_L h_{21}}V_2$$

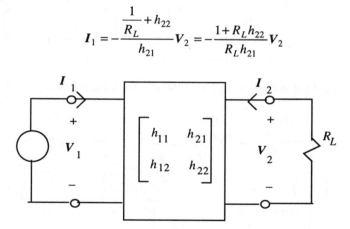

Figure 8.9. A configuration of an amplifier

From **Equation 8.2-13**, we have

$$V_1 = h_{11}I_1 + h_{12}V_2 = -\frac{h_{11}(1 + R_L h_{22})}{R_L h_{21}}V_2 + h_{12}V_2$$

Therefore, the amplification of the amplifier is found to be

$$A_v = \frac{V_2}{V_1} = \frac{V_2}{-\dfrac{h_{11}(1+R_L h_{22})}{R_L h_{21}}V_2 + h_{12}V_2} = \frac{R_L h_{21}}{-h_{11}(1+R_L h_{22})+h_{12}R_L h_{21}}$$

Problems

8.1-1. Find the voltage V_L of the circuit shown in **Figure 8.10**.

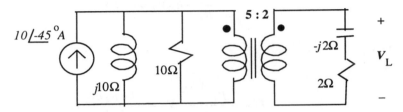

Figure 8.10. A circuit with a transformer

8.1-2. Find the voltage V_L in the circuit shown in **Figure 8.11**.

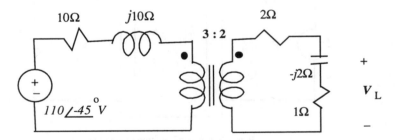

Figure 8.11. A circuit with a transformer

8.1-3. Find the current I_1 in the circuit shown in **Figure 8.12**.

8.1-4. Find the turn ratio N_2/N_1 for the ideal transformer in the circuit shown in **Figure 8.13** so that the maximum average power is transferred to the 8-Ω load. Find the average power delivered to the load.

8.2-1. Find the impedance parameters of the two-port network shown in **Figure 8.14**.

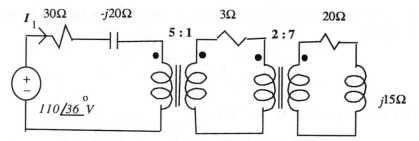

Figure 8.12. A circuit with two transformers

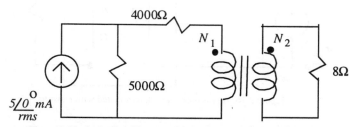

Figure 8.13. A circuit with a transformer for maximum power transfer

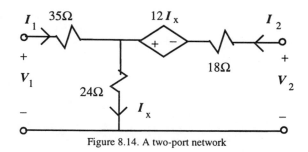

Figure 8.14. A two-port network

8.2-2. Find the h-parameters for the two-port shown in **Figure 8.15**.

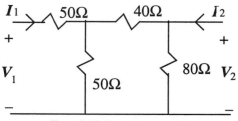

Figure 8.15. A two-port network

8.3-1. The transmission parameters for the two-port shown in **Figure 8.16** are given as follows:

$$T = \begin{bmatrix} 2 & 5 \\ 0.4 & 2 \end{bmatrix}$$

Find the power delivered to the 4-Ω load.

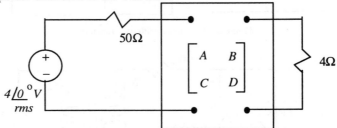

Figure 8.16. A two-port network terminated with resistors

Chapter 9
Fourier Analysis and Frequency Response

Some circuits are designed to handle many different kinds of signals. A constant signal (DC) and a sinusoidal signal with a fixed frequency (AC) are among just two of many types of signals encountered in such a circuit. These types of signals are special cases of the so-called periodic signals that are of particular interest to us. A periodic signal can be represented by a series, known as Fourier series, which consists of sinusoidal terms with various frequencies. The effect of each term in the series representation on a circuit can then be analyzed with the phasor technique that has been widely used in AC circuit analysis. The study of circuit responses due to sources with various frequency components is referred to as the frequency response. For some non-periodic signals, e.g., energy signals, Fourier transforms of the signals offer similar insights to the analysis of circuits.

9.1 Periodic Signals

A real-valued time function $x(t)$ is said to be periodic if there exist a positive time constant T_0 such that

$$x(t+T_0) = x(t) \qquad\qquad 9.1\text{-}1$$

for all time t.

The smallest positive time constant T_0 that makes **Equation 9.1-1** valid is known as the period of the function. For a periodic function, only the functional value over a particular period needs defining. For example a pulse train with a period of T_0 and duration d is defined as follows:

$$x(t) = \begin{cases} 1, & when & 0 \le t \le d \\ 0, & when & d < t < T_0 \\ x(t+mT_0), & otherwise & with \quad integer \quad m \end{cases} \qquad 9.1\text{-}2$$

The waveform of such a function is plotted in **Figure 9.1(a)**.

9.2 Fourier Series Representation

A periodic time function with a period T_0 can be represented by a series as in **Equation 9.2-1**.

$$x(t) = \sum_{k=0}^{\infty} a_k \cos k\omega_o t + \sum_{k=1}^{\infty} b_k \sin k\omega_o t \qquad 9.2\text{-}1$$

where the term $\omega_o = 2\pi / T_0$ is known as the fundamental angular frequency. The coefficients in **Equation 9.2-1** can be found as follows:

$$a_0 = \frac{1}{T_0} \int_0^{T_0} x(t)dt \qquad 9.2\text{-}2$$

$$a_k = \frac{2}{T_0} \int_0^{T_0} x(t)\cos(k\omega_o t)dt \qquad k = 1,2,3,..... \qquad 9.2\text{-}3$$

$$b_k = \frac{2}{T_0} \int_0^{T_0} x(t)\sin(k\omega_o t)dt \qquad k = 1,2,3,..... \qquad 9.2\text{-}4$$

With a minor adjustment, **Equation 9.2-1** can be converted into the following form:

$$x(t) = \sum_{k=0}^{\infty} \rho_k \cos(k\omega_o t - \theta_k) \qquad 9.2\text{-}5$$

It can be shown that

$$\rho_0 = a_0 \qquad with \quad \theta_0 = 0$$

and

$$\rho_k \cos \theta_k = a_k \quad and \quad \rho_k \sin \theta_k = b_k$$
$$with \quad k = 1,2,3,.....$$

The same periodic function can also be represented by the following series:

$$x(t) = \sum_{k=-\infty}^{\infty} c_k e^{jk\omega_o t} \qquad 9.2\text{-}6$$

The coefficients in **Equation 9.2-6** can be found as follows:

$$c_k = \frac{1}{T_0} \int_0^{T_0} x(t)e^{-jk\omega_o t} dt, \qquad k = 0,\pm1,\pm2,.... \qquad 9.2\text{-}7$$

Furthermore, we have

$$c_0 = \frac{1}{T_0} \int_0^{T_0} x(t)dt = a_0 = \rho_0 \qquad \text{9.2-8}$$

$$c_k = \frac{1}{T_0} \int_0^{T_0} x(t)e^{-jk\omega_o t}\, dt$$

$$= \frac{1}{T_0} \int_0^{T_0} x(t)[\cos(k\omega_o t) - j\sin(k\omega_o t)]dt = \frac{a_k - jb_k}{2} = \frac{\rho_k e^{-j\theta_k}}{2} \qquad \text{9.2-9}$$

when $k \neq 0$

$$c_{-k} = (c_k)^* = \left(\frac{a_k - jb_k}{2}\right)^* = \frac{a_k + jb_k}{2}, \quad k = 1,2,3,..... \qquad \text{9.2-10}$$

where the notation $(c_k)^*$ denotes the complex conjugate of the complex number "c_k".

For the pulse train represented by **Equation 9.1-2**, we have the following:

$$c_0 = \frac{1}{T_0} \int_0^{T_0} x(t)dt = \frac{1}{T_0} \int_0^d 1\, dt = \frac{d}{T_0} = a_0 = \rho_0$$

$$c_k = \frac{1}{T_0} \int_0^{T_0} x(t)e^{-jk\omega_o t}\, dt = \frac{1}{T_0} \int_0^d 1\, e^{-jk\omega_o t}\, dt = \frac{e^{-jk\omega_o t}}{-jT_0 k\omega_o}\bigg|_{t=0}^{d}$$

$$= \frac{1 - e^{-jk\omega_o d}}{jT_0 k\omega_o} = \frac{\sin(k\omega_o d)}{T_0 k\omega_o} - j\frac{1 - \cos(k\omega_o d)}{T_0 k\omega_o} \qquad \text{with} \quad k = 1,2,3,.....$$

Therefore, we have

$$a_k = \frac{2\sin(k\omega_o d)}{T_0 k\omega_o} = \frac{\sin(k\omega_o d)}{k\pi}$$

$$b_k = 2 \times \frac{1 - \cos(k\omega_o d)}{T_0 k\omega_o} = \frac{1 - \cos(k\omega_o d)}{k\pi} \qquad \text{with} \quad k = 1,2,3,.....$$

210

Furthermore, for $k \geq 1$,

$$|c_k| = \frac{\rho_k}{2} = \frac{\sqrt{a_k^2 + b_k^2}}{2} = \frac{2|\sin(k\omega_o d / 2)|}{T_0 k \omega_o} = \frac{d}{T_0} \times \frac{|\sin(\pi \xi_k)|}{\pi \xi_k}$$

$$= \frac{d}{T_0} |\text{sinc}(\xi_k)| \qquad \text{with} \quad \xi_k = \frac{k \omega_o d}{2\pi} = \frac{kd}{T_0}$$

where the function "sinc" is defined as follows:

$$\text{sinc}(x) = \frac{\sin(\pi x)}{\pi x}$$

It can further be shown that the pulse train has the following representation

$$x(t) = \frac{d}{T_0} + \frac{2d}{T_0} \sum_{k=1}^{\infty} \text{sinc}(\xi_k) \cdot \cos\left[k\omega_o\left(t - \frac{d}{2}\right)\right] \qquad 9.2\text{-}11$$

The term d/T_0 is known as the duty ratio.

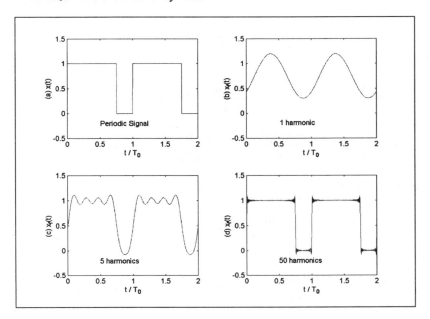

Figure 9.1 Expansion of a pulse train with duty ratio of ¾

Figure 9.1 demonstrates the convergence of the series toward the function $x(t)$ and validates its representation. Specifically, **Figure 9.1(a)** shows a pulse train with a duty ratio of 3/4. **Figure 9.1(b)** depicts the series involving only the first harmonic, whereas the waveform in **Figure 9.1(c)** represents the series containing the first 5 harmonics. Finally, the signal in **Figure 9.1(d)** is the sum of the first 50 harmonics. It is evident that the series converges to $x(t)$ as the number of harmonics tends to infinity.

The coefficient c_k in **Equation 9.2-6** can be associated with a function $C(\omega)$ of the angular frequency ω as follows:

$$C(\omega) = \begin{cases} c_k, & if \quad \omega = k\omega_o \\ 0, & otherwise \end{cases} \qquad \text{9.2-12}$$

The magnitude of $C(\omega)$ for the pulse train shown in **Figure 9.1(a)** can be plotted as in **Figure 9.2**. Since $C(\omega)$ is non-zero at only a set of countable points along the entire range of ω, it is known as the discrete spectrum of the given time function.

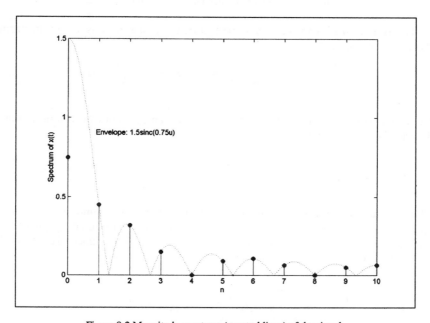

Figure 9.2 Magnitude spectrum (spectral lines) of the signal

212

9.3 Steady-State Analysis with Fourier Series

Since a periodic time function can be represented by a Fourier Series which is a linear combination of sinusoids with distinctive frequencies, steady-state analysis of a circuit driven by a periodic forcing function can be performed via linearity and the phasor technique as outlined in **Chapter 6**. To illustrate the procedures involved, let's examine the circuit in **Figure 9.3**.

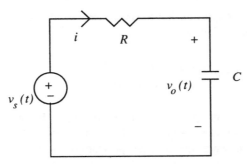

Figure 9.3 A first-order RC circuit driven by a periodic source

The source in **Figure 9.3** is assumed to be the pulse train as shown in **Figure 9.1(a)**. Hence, we have

$$v_s(t) = \sum_{k=0}^{\infty} \rho_k \cos(k\omega_o t - \theta_k) = \sum_{k=0}^{\infty} v_{sk}(t) \qquad 9.3\text{-}1$$

Due to linearity, the voltage across the capacitor can be found by assuming that it takes the following form:

$$v_o(t) = \sum_{k=0}^{\infty} v_{ok}(t) \qquad 9.3\text{-}2$$

Individual terms in the series can be found via the phasor technique as shown in **Figure 9.4**. Note that voltage terms $v_{sk}(t)$ and $v_{ok}(t)$ are represented by phasors V_{sk} and V_{ok}, respectively. The capacitor is represented by its impedance. The phsor V_{ok} is found as follows:

$$V_{ok} = \frac{1/(jk\omega_o C)}{R+1/(jk\omega_o C)} V_{sk} = \frac{1}{1+jk\omega_o CR} \rho_k e^{-j\theta_k}$$
$$= \frac{\rho_k}{\sqrt{1+(k\omega_o CR)^2}} e^{-j\left[\theta_k+\tan^{-1}(k\omega_o CR)\right]} \qquad 9.3\text{-}3$$

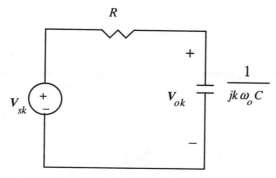

Figure 9.4 The equivalent circuit for phasor technique

Consequently, we have

$$v_{ok}(t) = \frac{\rho_k}{\sqrt{1+(k\omega_o CR)^2}} \cos\left[k\omega_o t - \theta_k - \tan^{-1}(k\omega_o CR)\right] \qquad 9.3\text{-}4$$

Figure 9.5 demonstrates the convergence of the series for the output signal in **Equation 9.3-2** toward the function $v_o(t)$ when the input signal is a pulse train. Specifically, **Figure 9.5(a)** shows an input pulse train with a duty ratio of 0.5. **Figure 9.5(b)** depicts the series of the output signal involving only the first two harmonics, whereas the waveform in **Figure 9.5(c)** represents the series containing the first 5 harmonics. Finally, the signal in **Figure 9.5(d)** is the sum of the first 50 harmonics.

9.4 Energy Signals and the Fourier Transform

The periodic signals covered in **Section 9.1** possess generally a finite amount of power. Theoretically, this type of signal possesses infinite amount of energy and thus cannot be generated artificially. A more realistic type of signal possesses only a finite amount of energy, thus the name energy signal. To deal with energy signals analytically, a different type of tool is needed. The Fourier transform is such a tool for analyzing energy signals.

A signal represented by the time function $f(t)$ is said to be an energy signal if its energy E_f, defined as in **Equation 9.4-1**, is finite.

$$E_f = \int_{-\infty}^{\infty} f^2(t)dt \qquad\qquad 9.4\text{-}1$$

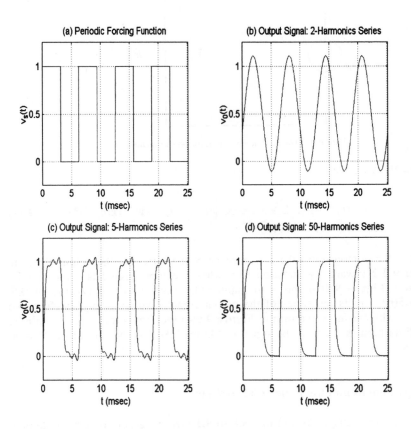

Figure 9.5 $T_0 = 2\pi$ (msec), duty ratio 0.5, and $RC = 0.05T_0$ (msec)

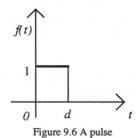

Figure 9.6 A pulse

A single pulse with duration d as shown in **Figure 9.6** is an energy signal. The energy of the signal $f(t)$ is found to be

$$E_f = \int_{-\infty}^{+\infty} f^2(t)dt = \int_0^d f^2(t)dt = \int_0^d 1dt = d \qquad 9.4\text{-}2$$

The Fourier Transform of an energy signal $f(t)$ is defined as follows:

$$F(\omega) = \int_{-\infty}^{+\infty} f(t) \cdot e^{-j\omega t} dt \qquad 9.4\text{-}3$$

For the pulse shown in **Figure 9.6**, we have

$$F(\omega) = \int_{-\infty}^{+\infty} f(t) e^{-j\omega t} dt = \int_0^d 1 e^{-j\omega t} dt = \frac{e^{-j\omega t}}{-j\omega} \bigg|_0^d = \frac{1 - e^{-j\omega \cdot d}}{j\omega}$$

$$= \frac{e^{-j\frac{\omega \cdot d}{2}}}{j\omega} \left(e^{j\frac{\omega \cdot d}{2}} - e^{-j\frac{\omega \cdot d}{2}} \right) = e^{-j\frac{\omega \cdot d}{2}} \times \frac{2 \sin \frac{\omega \cdot d}{2}}{\omega} \qquad 9.4\text{-}4$$

$$= e^{-j\frac{\omega \cdot d}{2}} \times d \times \mathrm{sinc}(\frac{\omega \cdot d}{2\pi})$$

In general, the Fourier transform of an energy signal is a complex valued function. The magnitude of the Fourier transform is often referred to as the spectrum of the signal. The spectrum of the pulse is of continuous nature as shown in **Figure 9.7**. Hence, it is also referred to as a continuous spectrum as opposed to the discrete spectrum of a periodic signal.

When a single pulse as in **Figure 9.6** is the source function for the circuit in **Figure 9.3**, the response of the circuit can be found via the phasor technique as was done in **Section 9.3**. The equivalent circuit for the phasor approach is shown in **Figure 9.8**. Note that the impedance of the capacitor is represented by $1/(j\omega C)$. Furthermore, the Fourier transform of the output signal is found as in **Equation 9.4-5**.

$$V_o = \frac{1/(j\omega C)}{R + 1/(j\omega C)} V_s = \frac{1}{1 + j\omega CR} F(\omega)$$

$$= \frac{1}{1+j\omega CR} \times e^{-j\frac{\omega \cdot d}{2}} \times \frac{2\sin\frac{\omega \cdot d}{2}}{\omega}$$

$$= \frac{2\sin\frac{\omega \cdot d}{2}}{\omega\sqrt{1+(\omega CR)^2}} \times e^{-j\left[\frac{\omega \cdot d}{2}+\tan^{-1}(\omega CR)\right]} \qquad \text{9.4-5}$$

The spectrum of the output signal is shown in **Figure 9.9**.

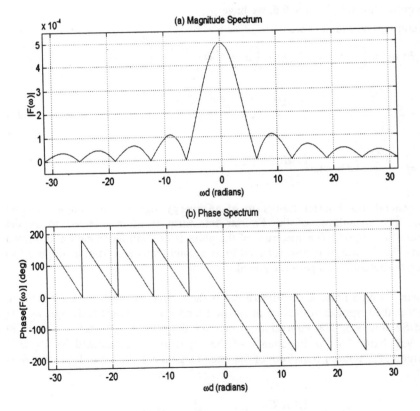

Figure 9.7 Spectrum of rectangular pulse $f(t)$ with $d = 0.5$ (msec)

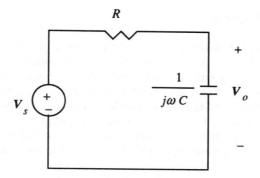

Figure 9.8. A first-order low-pass filter

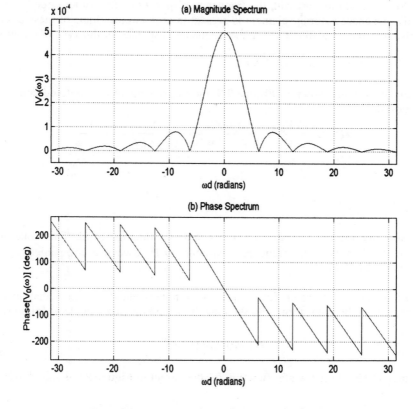

Figure 9.9 Spectrum of output pulse $v_o(t)$ with $RC = 0.1d = 0.05$ (msec)

9.5 Frequency Response

When a circuit is designed to respond to various frequency components, the characteristics of the circuit is normally frequency dependent. The characteristics of the circuit are generally represented through what are commonly referred to as the frequency responses. When the transfer function of a circuit is specified, the magnitude and the phase angle of the transfer function as functions of the frequency are of vital importance. In this section, we are going to discuss the frequency responses of the first-order, the second-order, and the higher-order circuits. The topics of Bode plots that are approximations of the frequency responses are also to be covered.

9.5.1 First-order Circuits

We are going to start with the frequency responses of first-order circuits where exactly one energy-storage element is present in each circuit. Some of these circuits are also referred to as filters due to the frequency responses of this type of circuits.

9.5.1.1 First-order Low-pass Filter

For the circuit shown in **Figure 9.8**, the transfer function is defined and found as follows:

$$H(j\omega) \overset{\text{def}}{=} \frac{V_o}{V_s} = \frac{1/(j\omega C)}{R + 1/(j\omega C)} = \frac{1}{1 + j\omega CR} = \frac{1}{1 + j\dfrac{\omega}{\omega_o}}$$

where $\omega_o = \dfrac{1}{CR}$.

Note that the transfer function is denoted as a function of $j\omega$. This notation is due to the fact that the frequency variable ω is always accompanied by the imaginary number j.

In general, a transfer function is a complex-valued function that can be represented in the polar form with the magnitude and the phase angle as follows:

$$H(j\omega) = M(\omega)\angle^{\phi(\omega)}$$

The magnitude of the transfer function for the circuit in **Figure 9.8** is found to be

$$M(\omega) = \frac{1}{\sqrt{1+(\omega/\omega_o)^2}}$$

The phase angle of the transfer function for the circuit in **Figure 9.8** is found to be

$$\phi(\omega) = -\tan^{-1}(\omega/\omega_o)$$

When the frequency ω is much smaller than ω_o, the magnitude of the transfer function can be approximated by 1. When the frequency ω is much greater than ω_o, the magnitude of the transfer function can be approximated by 0. A circuit with such characteristics is referred to as a low-pass filter. When the frequency ω is equal to ω_o, the square of the magnitude of the transfer function equals to ½ while the phase angle of the transfer function is – 45 degrees. This particular frequency is referred to as the half-power frequency.

9.5.1.2 First-order High-pass Filter

If we switch the positions of the capacitor and the resistor and assign the output voltage across the resistor, a high-pass filter is resulted as shown in **Figure 9.10**. For the circuit shown in **Figure 9.10**, the transfer function is defined and found as follows:

$$H(j\omega) \overset{\text{def}}{=} \frac{V_o}{V_s} = \frac{R}{R+1/(j\omega C)} = \frac{1}{1+1/(j\omega CR)} = \frac{1}{1-j\dfrac{\omega_o}{\omega}}$$

where $\omega_o = \frac{1}{CR}$.

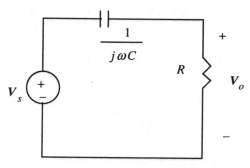

Figure 9.10 A first-order high-pass filter

The magnitude of the transfer function for the circuit in is found to be

$$M(\omega) = \frac{1}{\sqrt{1 + (\omega_o / \omega)^2}}$$

The phase angle of the transfer function for the circuit is found to be

$$\phi(\omega) = \tan^{-1}(\omega_o / \omega)$$

When the frequency ω is much smaller than ω_o, the magnitude of the transfer function can be approximated by 0. When the frequency ω is much greater than ω_o, the magnitude of the transfer function can be approximated by 1. A circuit with such characteristics is referred to as a high-pass filter. When the frequency ω is equal to ω_o, the square of the magnitude of the transfer function equals to ½ while the phase angle of the transfer function equals to 45 degrees.

9.5.2 Second-order Circuits

A second-order circuit contains two energy storage elements. Among all the second-order circuits, of particular interests to us are the so-called resonant circuits.

9.5.2.1 Series Resonant Circuit

For the circuit shown in **Figure 9.11**, a transfer function is defined and found as follows:

$$H(j\omega) = \frac{V_o}{V_i} = \frac{R}{R + j\omega L + 1/(j\omega C)} = \frac{1}{1 + j(\dfrac{\omega L}{R} - \dfrac{1}{R\omega C})}$$

The magnitude of the transfer function is found to be

$$M(\omega) = \frac{1}{\sqrt{1 + (\dfrac{\omega L}{R} - \dfrac{1}{R\omega C})^2}}$$

The phase angle of the transfer function is found to be

$$\phi(\omega) = -\tan^{-1}(\frac{\omega L}{R} - \frac{1}{R\omega C})$$

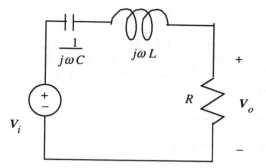

Figure 9.11 A series resonant circuit

The magnitude of the transfer function equal to 1 when the frequency ω is such that

$$\frac{\omega L}{R} - \frac{1}{R\omega C} = 0$$

i.e., when

$$\omega = \frac{1}{\sqrt{LC}}$$

This frequency is known as the resonant frequency ω_r. Note that the phase angle of the transfer function is of zero degree at the resonant frequency. Hence, the output voltage and the input voltage are in phase at the resonant frequency.

The frequencies at which the square of the magnitude of the transfer function becomes ½ are known as the half-power frequencies. These frequency can be found if we set

$$\frac{\omega L}{R} - \frac{1}{R\omega C} = \pm 1 \qquad\qquad 9.5\text{-}1$$

There are two solutions that are positive real for **Equation 9.5-1**. They are identified as ω_1 and ω_2, in ascending order. These two frequencies are related in the following way:

$$\omega_1\omega_2 = \frac{1}{LC} = \omega_r{}^2$$

$$\omega_2 - \omega_1 = R/L$$

When the frequency approaches either to zero or to infinity, the magnitude of the transfer function approaches to zero. Circuits with such characteristics are known as band-pass filters. The difference of the two half-power frequencies as shown in the equation above is also known as the bandwidth of the circuit.

9.5.2.2 Parallel Resonant Circuit

For the circuit shown in **Figure 9.12**, a transfer function is defined and found as follows:

$$H(j\omega) \overset{def}{=} \frac{I_o}{I_i} = \frac{1/R}{j\omega C + 1/(j\omega L) + 1/R} = \frac{1}{1 + j(\omega CR - \dfrac{R}{\omega L})}$$

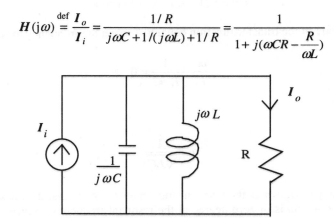

Figure 9.12 A parallel resonant circuit

The magnitude of the transfer function is found to be

$$M(\omega) = \frac{1}{\sqrt{1 + (\omega CR - \dfrac{R}{\omega L})^2}}$$

The phase angle of the transfer function is found to be

$$\phi(\omega) = -\tan^{-1}(\omega CR - \frac{R}{\omega L})$$

The magnitude of the transfer function equal to 1 when the frequency ω is such that

$$\omega CR - \frac{R}{\omega L} = 0$$

i.e., when

$$\omega = \frac{1}{\sqrt{LC}}$$

This frequency is known as the resonant frequency ω_r. Note that the resonant frequency has the same expression whether it is the series or the parallel resonant circuit.

The half-power frequencies of the parallel resonant circuit can be found if we set

$$\omega CR - \frac{R}{\omega L} = \pm 1 \qquad\qquad 9.5\text{-}3$$

There are two solutions that are positive real for **Equation 9.5-3**. They are also identified as ω_1 and ω_2, in ascending order. These two frequencies are found to be with the following properties:

$$\begin{cases} \omega_1\omega_2 = \dfrac{1}{LC} = \omega_r{}^2 \\[2mm] \omega_2 - \omega_1 = \dfrac{1}{CR} \end{cases}$$

When the frequency approaches either to zero or to infinity, the magnitude of the transfer function approaches to zero. Hence, the circuit is also a band-pass filter.

9.5.3 Complex Frequency and General Circuit Analysis

For a third- or a higher-order circuit, the analytic expressions of transfer functions become more complex. Fortunately, the frequency variable ω is always accompanied by the imaginary number j in a product for the expression of a transfer function as noted. This product $j\omega$ can be represented by a complex number s that is often referred to as the complex frequency. With the complex frequency replacing the product, the impedance of an inductor is represented as sL and the impedance of a capacitor is represented as $1/(sC)$. Consequently, a transfer function of a circuit can be derived via circuit laws or network analysis techniques and be proven to be a rational function of the complex frequency with real coefficients as follow:

$$H(s) = \frac{b_m s^m + b_{m-1} s^{m-1} + \dots + b_1 s + b_0}{s^n + a_{n-1} s^{n-1} + \dots + a_1 s + a_0}$$

9.5-4

As polynomials of the complex frequency, both the numerator and the denominator of **Equation 9.5-4** can be factorized into terms represented by the roots of the polynomials as follows:

$$H(s) = \frac{b_m (s - z_1)(s - z_2) \dots (s - z_m)}{(s - p_1)(s - p_2) \dots (s - p_n)}$$

9.5-5

The roots of the numerator are known as the zeros of the transfer function while the roots of the denominator are known as the poles of the transfer function.

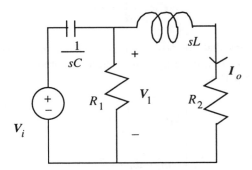

Figure 9.13 A RLC circuit

Example 9.1 The transfer function of the circuit shown in **Figure 9.13** is defined as the ratio of I_o and V_i. Find the transfer function.

Solution: Via KCL, we have

$$sC(V_1 - V_i) + \frac{V_1}{R_1} + \frac{V_1}{sL + R_2} = 0$$

Solving for V_1, we have

$$V_1 = \frac{sCV_i}{sC + \dfrac{1}{R_1} + \dfrac{1}{sL+R_2}} = \frac{sCR_1(sL+R_2)}{sCR_1(sL+R_2)+(sL+R_2)+R_1} V_i$$

Consequently, the transfer function is found to be

$$H(s) \stackrel{\text{def}}{=} \frac{I_o}{V_i} = \frac{I_o}{V_1} \cdot \frac{V_1}{V_i} = \frac{1}{(sL+R_2)} \cdot \frac{sCR_1(sL+R_2)}{sCR_1(sL+R_2)+(sL+R_2)+R_1}$$

$$= \frac{sCR_1}{sCR_1(sL+R_2)+(sL+R_2)+R_1} = \frac{sCR_1}{CLR_1 s^2 + (CR_1R_2 + L)s + (R_2+R_1)}$$

9.5.4 Bode Plots

From **Equation 9.5-5**, the frequency responses of a circuit can be found by simply substituting the complex frequency s with the product $j\omega$ as shown in the following equation.

$$H(j\omega) = \frac{b_m(j\omega - z_1)(j\omega - z_2).....(j\omega - z_m)}{(j\omega - p_1)(j\omega - p_2).....(j\omega - p_n)} \qquad 9.5\text{-}6$$

Without loss of generality, **Equation 9.5-6** can also be represented as follows:

$$H(j\omega) = \frac{K(j\omega)^q \left(1 - \dfrac{j\omega}{\tilde{z}_1}\right) \cdot \left(1 - \dfrac{j\omega}{\tilde{z}_2}\right).... \left(1 - \dfrac{j\omega}{\tilde{z}_\mu}\right)}{\left(1 - \dfrac{j\omega}{\tilde{p}_1}\right) \cdot \left(1 - \dfrac{j\omega}{\tilde{p}_2}\right).... \left(1 - \dfrac{j\omega}{\tilde{p}_v}\right)} \qquad 9.5\text{-}7$$

The factor $(j\omega)^q$ reflects the presence of zeros or poles at the origin of the complex plane.

When the transfer function is represented by the product of two or more parts, the phase angle of the transfer function equals to the sum of the phase angles of the individual parts. This property is illustrated as follows:

Let

$$H(j\omega) = \prod_{k=1}^{n} H_k(j\omega) = H_1(j\omega)H_2(j\omega)\cdots H_n(j\omega)$$

where

$$H_k(j\omega) = M_k(\omega)\angle^{\phi_k(\omega)}, \quad k = 1,2,...,n$$

We have

$$\phi(\omega) = \sum_{k=1}^{n} \phi_k(\omega) \qquad\qquad 9.5\text{-}8$$

and

$$M(\omega) = \prod_{k=1}^{n} M_k(\omega)$$

As shown in the equations above, the magnitude of a transfer function does not have the same property as that of the phase angle. However, taking the log of the magnitude could turn multiplication into summation. Consequently, the magnitude of a transfer function is often measured in decibel (or *dB*) which is defined as follows:

$$20 \cdot \log[M(\omega)] = 20 \cdot \log\left[\prod_{k=1}^{n} M_k(\omega)\right] = \sum_{k=1}^{n} 20 \cdot \log[M_k(\omega)] \qquad 9.5\text{-}9$$

With **Equations 9.5-8** and **9.5-9**, the magnitude (in *dB*) and the phase angle of the transfer function can be found by first studying the contributions by individual terms shown in **Equation 9.5-7**. These individual terms are categorized in the following cases:

    (1) the constant term
    (2) the $j\omega$ term
    (3) the first-order term with a corner frequency
    (4) the second-order term

Using the log scale for the frequency ω, the magnitude plot (in *dB*) and the phase angle plot for **Case (1)** or **(2)** are straight-line diagrams. Although the magnitude plot (in *dB*) and the phase angle plot for **Case (3)** or **(4)** are not straight-line diagrams, they can be approximated with segments of straight-lines. The approximations of frequency responses with straight-line segments are known as Bode plots. The Bode plots of various cases are discussed in the following.

Case 1: The constant term

Since the constant term is frequency independent, the magnitude plot or the phase angle plot is a straight-line with zero slope as shown in **Figure 9.14**.

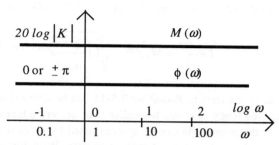

Figure 9.14 The magnitude and the phase angle of a constant

Note that the angle of a positive constant is 0 while the angle of a negative constant is $\pm \pi$ (or $\pm 180°$).

Case 2: The $j\omega$ term

The magnitude of the $j\omega$ term is given as follows:

$$20\log\left|(j\omega)^q\right| = 20q\log\left|j\omega\right| = 20q\log\omega \qquad 9.5\text{-}10$$

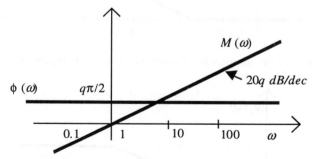

Figure 9.15 The magnitude and the phase angle of the $j\omega$ term

Equation 9.5-10 represents a straight line with a slope of 20q *dB/decade* as shown in **Figure 9.15**. The slope is positive if the power q is positive and is negative if q is negative. A decade represents an increase in frequency by a factor of 10. Also shown in **Figure 9.15** is the phase angle of the $j\omega$ term, which is constant.

Case 3: The first-order term with a corner frequency

A first-order term due to a negative real zero takes the following form:

$$1 + \frac{j\omega}{\omega_o} = M_1 \angle^{\phi_1} = \sqrt{1 + \frac{\omega^2}{\omega_o^2}} \angle^{tan^{-1}(\omega/\omega_o)} \qquad 9.5\text{-}11$$

The magnitude of the term in **Equation 9.5-11** can be approximated by 1 (i.e. 0 dB) when the frequency ω is much smaller than ω_o, say a tenth or smaller. On the other hand, the magnitude can be approximated by ω/ω_o if the frequency ω is much greater than ω_o, say ten times or greater. With log scale, these two approximations are of straight-line characteristics as in **Cases 1** and **2**. The two straight lines intersect at the frequency ω_o, which is known as the corner frequency. The approximation for the magnitude is shown in **Figure 9.16**. It should be noted that the maximum error of the approximation is about 3 dB, which occurs right at the corner frequency.

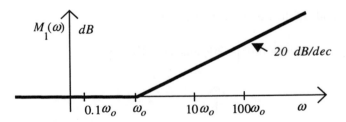

Figure 9.16 The magnitude plot of a first-order term due to a negative real zero

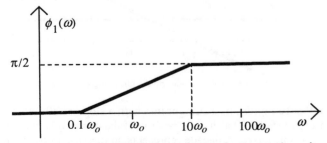

Figure 9.17 The phase anlgle plot of a first-order term due to a negative real zero

The phase angle of the term in **Equation 9.5-11** can be approximated by 0 when the frequency ω is much smaller than ω_o. If the frequency ω is much greater than ω_o, the phase angle can be approximated by $\pi/2$ (or 90 degrees.) When the frequency ω is equal to ω_o, the phase angle is exactly $\pi/4$ (or 45 degrees.) The approximation for the phase angle is shown in **Figure 9.17**.

The Bode plots of the first-order term due to a negative real pole are shown in **Figures 9.18** and **9.19**. These plots are negative to the plots in **Figures 9.16** and **9.17** due to the fact that a pole came from the denominator of the transfer function.

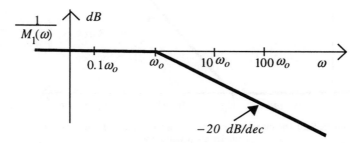

Figure 9.18 The magnitude plot of a first-order term due to a negative real pole

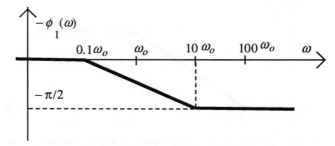

Figure 9.19 The phase angle plot of a first-order term due to a negative real pole

Case 4: The second-order term

A second-order term due to a pair of complex-conjugate zeros takes the following form:

$$1 + \frac{j2\xi\omega}{\omega_o} - \frac{\omega^2}{\omega_o{}^2} = M_2(\omega)\angle^{\phi_2(\omega)} \qquad 9.5\text{-}12$$

where

$$M_2(\omega) = \sqrt{\left(1 - \frac{\omega^2}{\omega_o^2}\right)^2 + \left(\frac{2\xi\omega}{\omega_o}\right)^2} \quad and \quad \phi_2(\omega) = \tan^{-1}\left(\frac{2\xi\omega/\omega_o}{1 - \frac{\omega^2}{\omega_o^2}}\right)$$

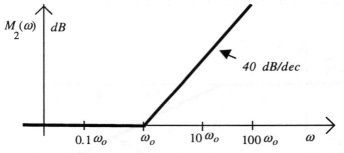

Figure 9.20 The magnitude plot of a second-order term due to a pair of complex conjugate zeros

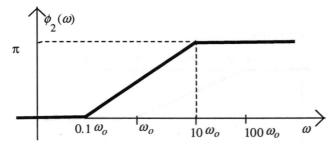

Figure 9.21 The phase angle plot of a second-order term due to a pair of complex conjugate zeros

The magnitude of the term in **Equation 9.5-12** can be approximated by 1 (i.e. 0 dB) when the frequency ω is much smaller than ω_o. On the other hand, the magnitude can be approximated by $(\omega/\omega_o)^2$ if the frequency ω is much greater than ω_o. With log scale, these two approximations are also of straight-line characteristics as in **Cases 1** and **2**. The two straight lines intersect at the frequency ω_o. The Bode plot for the magnitude is shown in **Figure 9.20**. Note that the slope of the straight-line segment for high frequency is of 40 dB/decade. It should also be noted that the maximum error of the approximation is dependent on the damping ratio ξ. The range of the damping ratio is between 0 and 1. In general, the smaller damping ratio the larger the maximum error.

The phase angle of the term in **Equation 9.5-12** can be approximated by 0 when the frequency ω is much smaller than ω_o. If the frequency ω is much greater than ω_o, the phase angle can be approximated by π (or 180 degrees.) When the frequency ω is equal to ω_o, the phase angle is exactly $\pi/2$ (or 90 degrees.) The approximation for the phase angle is shown in **Figure 9.21.**

The Bode plots of the first-order term due to a pair of complex-conjugate poles are shown in **Figures 9.22** and **9.23**. These plots are negative to the plots in **Figures 9.20** and **9.21** due to the fact that a pole came from the denominator of the transfer function.

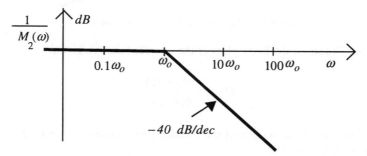

Figure 9.22 The magnitude plot of a second-order term due to a pair of complex conjugate poles

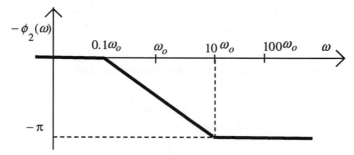

Figure 9.23 The phase angle plot of a second-order term due to a pair of complex conjugate poles

Once the individual Bode plots of the components of a transfer function are plotted, the Bode plots of a transfer function can be constructed by adding parts

232

together. The adding of parts is best carried out from the left to the right. This process is demonstrated with the **Example 9.2**.

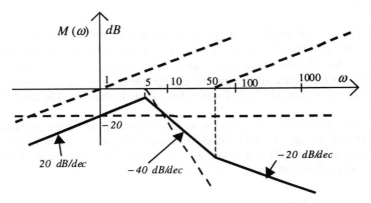

Figure 9.24 The magnitude plots for Example 9.2

Example 9.2 Construct the Bode plots of the transfer function given.

$$H(s) = \frac{s(s+50)}{4(s+5)^3}$$

Solution: First, we need to convert the transfer function into the form that leads to **Equation 9.5-7** as follows:

$$H(s) = \frac{s(\frac{s}{50}+1)\times 50}{4(\frac{s}{5}+1)^3 \times 5^3} = \frac{0.1\times s(\frac{s}{50}+1)}{(\frac{s}{5}+1)^3}$$

For the Bode plots, we replace the complex frequency s with $j\omega$.

$$H(j\omega) = \frac{0.1\times j\omega(\frac{j\omega}{50}+1)}{(\frac{j\omega}{5}+1)^3}$$

There are four components in the transfer function. They are a constant term, a $j\omega$ term, a first-order term with a corner frequency of 50 *radians/s*, and a triple of first-order terms with the same corner frequency of 5 *radians/s*. The magnitude plots for the components and the transfer function are shown in **Figure 9.24**. The phase angle plots for the components and for the transfer function are shown in **Figure 9.25**.

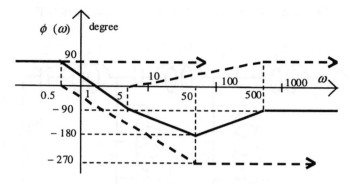

Figure 9.25 The phase angle plots for Example 9.2

Problems

9.2-1 Calculate the Fourier series of the function shown in **Figure 9.26**.

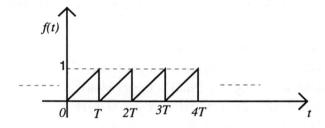

Figure 9.26 A saw-tooth waveform

9.2-2 Calculate the Fourier series of the following function

$$f(t) = 10|\sin(2t)|$$

9.5-1 For the circuit shown in **Figure 9.27**, use a PSpice program (or an equivalent one) to plot the magnitude (in dB) and the phase angle of the transfer function $H(j\omega)=V_2/V_1$ as functions of frequency ω (log scale). In the circuit, V_1 represents a sinusoidal input of variable frequency ω. The parameters of the elements are as follows:

$R_1 = R_2 = 1\Omega$

$L_1 = L_5 = 67.874\ \mu H$ $L_2 = L_4 = 16.180\ \mu H$ $L_3 = 101.110\ \mu H$

$C_1 = C_5 = 11.662\ \mu F$ $C_2 = C_4 = 48.923\ \mu F$ $C_3 = 7.8297\ \mu F$

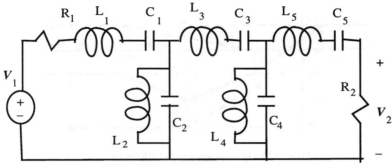

Figure 9.27 A ladder network

9.5-2. For the circuit shown in **Figure 9.28**, use a PSpice program (or an equivalent one) to plot the magnitude (in dB) and the phase angle of the transfer function $H(j\omega) = V_2/V_1$ as functions of frequency ω (log scale). In the circuit, V_1 represents a sinusoidal input of variable frequency ω. The parameters of the elements are as follows:

$R_1 = 1K\Omega$ $R_2 = 2.65974K\Omega$

$L_1 = 342.912mH$ $L_2 = 7.2078mH$ $L_3 = 487.56mH$

$L_4 = 6.9096mh$ $L_5 = 467.39mH$ $L_6 = 9.824mH$

$C_1 = 3.69363nF$ $C_2 = 175.73nF$ $C_3 = 2.59787nF$

$C_4 = 183.31nF$ $C_5 = 2.71nF$ $C_6 = 128.93nF$

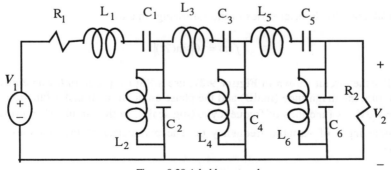

Figure 9.28 A ladder network

9.5-3 Draw the magnitude characteristic of the Bode plot for the transfer function

$$H(j\omega) = \frac{100(j\omega)^2(j\omega+50)}{(j\omega+5)^2(j\omega+100)}$$

9.5-4 Draw the magnitude characteristic of the Bode plot for the transfer function

$$H(j\omega) = \frac{10(j\omega+1)(j\omega+20)}{(j\omega)^2(j\omega+100)}$$

9.5-5 Draw the magnitude characteristic of the Bode plot for the transfer function

$$H(j\omega) = \frac{(j\omega+1)(j\omega+200)}{5(j\omega)^2(j\omega+10)}$$

9.5-5 Draw the magnitude characteristic of the Bode plot for the transfer function given.

$$H(s) = \frac{s(s+20)}{(s+1)^2(s+200)}$$

9.5-6 Find the transfer function $H(j\omega)$ if its magnitude characteristics is as shown in **Figure 9.29**.

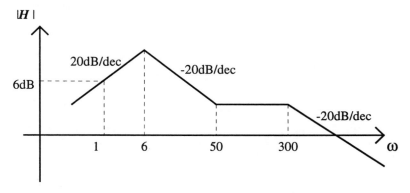

Figure 9.29 The Bode plot of a transfer function

9.5-7 Find the transfer function $H(j\omega)$ if its magnitude characteristics is as shown in **Figure 9.30**.

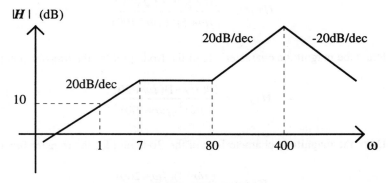

Figure 9.30 The Bode plot of a transfer function

Chapter 10
Laplace Transform

To deal with signals that may be energy signals, power signals, or combinations of both, the Laplace transform is more desirable than the Fourier transform. Another advantage of Laplace transform is that the transform of a real-valued signal has an expression with real coefficients.

10.1 Definition

Let $f(t)$ be a function that maps from $[0, \infty)$ to the set of complex numbers C. The Laplace Transform of $f(t)$ is defined according to the following equation

$$L\{f(t)\} = \int_0^\infty f(t)e^{-st}\,dt = F(s) \qquad\qquad 10.1\text{-}1$$

if the integral in the equation exists for some real number $s = \lambda$. The variable s has the units of frequency and is allowed to be a complex number. Hence, s is referred to as the complex frequency. The set of all the complex numbers at which the Laplace integral converges is known as the region of convergence. Let us examine the following example.

Example 10.1 Let $f(t)$ be a constant function, i.e., $f(t) = K$ for $t \geq 0$. Find the Laplace transform of the function.

Solution: From the definition, we have

$$F(s) = \int_0^\infty Ke^{-st}\,dt = -\frac{K}{s}e^{-st}\bigg|_0^\infty = \frac{K}{s}$$

if $s > 0$ or if the real part of $s = Re\{s\} > 0$.

Example 10.2 Let $f(t)$ be an exponential function, i.e., $f(t) = Ke^{at}$ for $t \geq 0$. Find the Laplace transform of the function.

Solution: We have

$$F(s) = \int_0^\infty Ke^{at}e^{-st}\,dt = -\frac{K}{s-a}e^{-(s-a)t}\bigg|_{t=0}^\infty = \frac{K}{s-a}$$

if $s > a$ or if $Re\{s\} > Re\{a\}$.

For our purpose, two functions that have the same Laplace transform are considered the same. Consequently, we write

$$f(t) = L^{-1}\{F(s)\}$$

where L^{-1} denotes the inverse Laplace transform. The method of finding $f(t)$ from $F(s)$ is deferred until we finish discussing the characteristics of the Laplace transform.

10.2. Properties of Laplace Transform

The Laplace transform and the inverse Laplace transform can be considered as operators that map from a function space to another function space. These operators possess desirable properties such as linearity. With these properties, the Laplace transform and the inverse Laplace transform are suitable for handling ordinary differential equations that characterize linear RLC circuits.

10.2.1 Linearity

Let $f(t)$ and $g(t)$ be functions from $[0,\infty)$ to C with Laplace transform $F(s)$ and $G(s)$, respectively. Let a and b be complex numbers. We have

$$L\{af(t) + bg(t)\} = aL\{f(t)\} + bL\{g(t)\} = aF(s) + bG(s) \qquad 10.2\text{-}1$$

Similarly, the inverse Laplace transform L^{-1} is also linear.

Example 10.1 Let $f(t)$ be the cosine function, i.e., $f(t) = \cos(\omega t)$. Find the Laplace transform of $f(t)$.

Solution: Using the linearity of the Laplace transform, we have

$$F(s) = L\{\cos(\omega \cdot t)\} = L\{\tfrac{1}{2}e^{j\omega \cdot t} + \tfrac{1}{2}e^{-j\omega \cdot t}\} = \tfrac{1}{2}L\{e^{j\omega \cdot t}\} + \tfrac{1}{2}L\{e^{-j\omega \cdot t}\}$$

$$= \frac{1}{2(s - j\omega)} + \frac{1}{2(s + j\omega)} = \frac{(s + j\omega) + (s - j\omega)}{2(s - j\omega)(s + j\omega)} = \frac{s}{s^2 + \omega^2}$$

10.2.2 Frequency Shifting

When a function $f(t)$ is multiplied with an exponential function $e^{\sigma t}$, the Laplace transform of the product $f(t)e^{\sigma t}$ can be found by replacing s in $L\{f(t)\}=F(s)$ with

$(s\text{-}\sigma)$. The proof is as shown in **Equation 10.2-2**.

$$L\{f(t)e^{\sigma t}\} = \int_0^\infty f(t)e^{\sigma t}e^{-st}\,dt = \int_0^\infty f(t)e^{-(s-\sigma)t}\,dt = F(s-\sigma) \qquad 10.2\text{-}2$$

Example 10.2 Find the Laplace transform of $e^{\sigma t}\cos(\omega t)$.

Solution: With the frequency shifting, we have

$$L\{e^{\sigma t}\cos(\omega \cdot t)\} = L\{\cos(\omega \cdot t)\}\big|_{s=s-\sigma} = \frac{s}{s^2+\omega^2}\bigg|_{s=s-\sigma} = \frac{s-\sigma}{(s-\sigma)^2+\omega^2}$$

10.2.3 Time Shifting

Let $u(t\text{-}\tau)$ denote the function from $[0,\infty)$ to C with the following properties.

$$u(t-\tau) = \begin{cases} 1, & when & t \geq \tau > 0 \\ 0, & when & 0 \leq t < \tau \end{cases}$$

We have

$$L\{f(t-\tau)u(t-\tau)\} = \int_0^\infty f(t-\tau)u(t-\tau)e^{-st}\,dt = \int_\tau^\infty f(t-\tau)e^{-st}\,dt$$

$$= \int_0^\infty f(x)e^{-s(x+\tau)}\,dt = e^{-s\tau}\int_0^\infty f(x)e^{-sx}\,dt = e^{-s\tau}F(s) \qquad 10.2\text{-}3$$

10.2.4 Differentiation

Assume that the function $f(t)$ has Laplace transform $F(s)$ and the derivative of $f(t)$ exists almost everywhere. If the derivative of $f(t)$, denoted as $f'(t)$, is piecewise continuous, then the Laplace transform of $f'(t)$ can be found as in the following equation:

$$L\{f'(t)\} = \int_0^\infty f'(t)e^{-st}\,dt$$

$$= f(t)e^{-st}\big|_{t=0}^\infty - \int_0^\infty f(t)[-se^{-st}]\,dt = -f(0) + sF(s) \qquad 10.2\text{-}4$$

In the derivation, we have used the integration by parts and have applied the fact that

$$\lim_{t \to \infty} f(t)e^{-st} = 0$$

which is a necessary condition for the existence of $L\{f(t)\}$.

If the second derivative of f(t), denoted as f'', is piecewise continuous, we then have

$$L\{f''(t)\} = \int_0^\infty f''(t)e^{-st}dt = \int_0^\infty [f'(t)]'e^{-st}dt = -f'(0) + sL\{f'(t)\}$$
$$= -f'(0) + s[-f(0) + sF(s)] = -f'(0) - sf(0) + s^2 F(s)$$

In general, we have

$$L\{f^{(n)}(t)\} = s^n F(s) - s^{n-1} f(0) - s^{n-2} f'(t) - \ldots - sf^{(n-2)}(0) - f^{(n-1)}(t)$$

where $f^{(n)}(t)$ denotes the n-th derivative of f(t) and is assumed to be piecewise continuous.

10.2.5 Integration

Assume that the function f(t) has Laplace transform F(s). The Laplace transform of the integration of f(t) can be found as follows:

$$L\{f(t)\} = L\{[\int_{-\infty}^t f(t)dt]'\} = sL\{\int_{-\infty}^t f(t)dt\} - \int_{-\infty}^0 f(t)dt$$

Hence,

$$L\{\int_{-\infty}^t f(t)dt\} = \frac{1}{s}L\{f(t)\} + \frac{1}{s}\int_{-\infty}^0 f(t)dt = \frac{1}{s}F(s) + \frac{1}{s}\int_{-\infty}^0 f(t)dt \quad 10.2\text{-}5$$

With the properties of Laplace transform discussed above, we are ready to analyze RLC circuits. Let us examine the circuit in the following example.

Example 10.3 Assuming that i(0) and v(0) equal to zero. Find the current i(t) for t≥0 of the circuit in **Figure 10.1**.

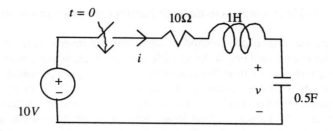

Figure 10.1. An RLC circuit

Solution: Applying Kirchoff's voltage law to the circuit for $t \geq 0$, we have

$$10 = 3i(t) + 1\frac{di}{dt} + v(t) \qquad \text{10.2-6}$$

The voltage across the capacitor is related to the current as follows:

$$v(t) = \frac{1}{0.5}\int_{-\infty}^{t} i(t)dt$$

Taking the Laplace transform of both sides of **Equation 10.2-6**, we have

$$\frac{10}{s} = 3I(s) + [sI(s) - i(0)] + V(s)$$

$$= 3I(s) + [sI(s) - i(0)] + 2[\frac{1}{s}I(s) + \frac{1}{s}\int_{-\infty}^{0} i(t)dt]$$

$$= 3I(s) + [sI(s) - i(0)] + \frac{2}{s}I(s) + \frac{1}{s}v(0) = 3I(s) + sI(s) + \frac{2}{s}I(s)$$

Solving for $I(s)$, we have

$$I(s) = \frac{10/s}{3 + s + 2/s} = \frac{10}{s^2 + 3s + 2}$$

The current $i(t)$ is found by taking the inverse Laplace transform of $I(s)$. The procedures will be outlined in the following section.

10.3 Inverse Laplace Transform through Partial Fraction Expansion

There are more than one method to carry out the inverse Laplace transform. The most often-mentioned one is the so-called partial fraction expansion. In general, the Laplace transform of a time function consists of strictly proper rational functions or their products with exponential terms such as e^{st}. Such an exponential term is normally caused by some kind of delay of certain time function. It can usually be processed separately. Hence, without losing generality, we assume that the Laplace transform of a time function takes the form of a strictly proper rational function. In other words, we have

$$F(s) = \frac{c(s)}{d(s)}$$

where $c(s)$ and $d(s)$ are polynomials with the degree of $c(s)$ less than the degree of $d(s)$. The roots of $d(s)$ are known as the poles of $F(s)$ while the roots of $c(s)$ are known as zeros of $F(s)$. In order to carry out the partial fraction expansion, we need to find the poles of $F(s)$ first. Assume that the degree of $d(s)$ is n. The poles of $F(s)$ are denoted as $p_1, p_2, ..., p_n$. The locations of these poles affect the procedure. There are basically two cases: distinctive poles and repeated poles. These two cases are treated separately in the following sections.

10.3.1 Distinctive Poles

If no two poles are the same, the Laplace transform $F(s)$ can be expressed as follows:

$$F(s) = \frac{a_1}{s - p_1} + \frac{a_2}{s - p_2} + ... + \frac{a_n}{s - p_n}$$

The coefficients are found by calculating the following limits

$$a_i = \lim_{s \to p_i} (s - p_i)F(s)$$

where i varies from 1 through n.

Example 10.3 (continued) Find the current $i(t)$ for $t \geq 0$ of the circuit in **Figure 10.1**.

Solution: Taking the partial fraction expansion of the Laplace transform found previously, we have

$$I(s) = \frac{10}{s^2 + 3s + 2} = \frac{10}{(s+1)(s+2)} = \frac{a}{s+1} + \frac{b}{s+2}$$

where

$$a = \lim_{s \to -1} (s+1)I(s) = \frac{10}{s+2}\Big|_{s=-1} = \frac{10}{-1+2} = 10$$

and

$$b = \lim_{s \to -2} (s+2)I(s) = \frac{10}{s+1}\Big|_{s=-2} = \frac{10}{-2+1} = -10$$

Consequently,

$$i(t) = L^{-1}\{I(s)\} = L^{-1}\{\frac{10}{s+1}\} - L^{-1}\{\frac{10}{s+2}\} = 10e^{-t} - 10e^{-2t}$$

for $t > 0$.

10.3.2 Repeated Poles

Without losing generality, let us assume that the number p_1 is the only repeated pole and is repeated m times. In other words, the denominator of the Laplace transform $F(s)$ can be factorized as $(s\text{-}p_1)^m (s\text{-}p_{m+1})\cdots(s\text{-}p_n)$. The Laplace transform $F(s)$ may be expressed as follows:

$$F(s) = \frac{a_1}{s - p_1} + \frac{a_2}{(s - p_1)^2} + \ldots + \frac{a_m}{(s - p_1)^m} + \frac{a_{m+1}}{s - p_{m+1}} + \ldots + \frac{a_n}{s - p_n}$$

The coefficients from a_{m+1} through a_n can be found through the same procedure as in **Section 10.3.1**. The coefficients a_1 through a_m are to be calculated as in the following. From the equation above, we have

$$(s - p_1)^m F(s) = a_1 (s - p_1)^{m-1} + \ldots + a_{m-1}(s - p_1) + a_m$$

$$+ \frac{a_{m+1}(s - p_1)^m}{s - p_{m+1}} + \ldots + \frac{a_n (s - p_1)^m}{s - p_n}$$

Taking the limit as s approaches p_1, we have

$$a_m = \lim_{s \to p_1} (s - p_1)^m F(s)$$

To find a_{m-1}, we take the derivative of $(s-p_1)^m F(s)$ first, then we find the limit as s approaches p_1. We thus have

$$a_{m-1} = \lim_{s \to p_1} \frac{d}{ds} \{(s - p_1)^m F(s)\}$$

In general, we have

$$(k!) \cdot a_{m-k} = \lim_{s \to p_1} \frac{d^k}{ds^k} \{(s - p_1)^m F(s)\}$$

where k varies from 1 to $(m-1)$. By the way, a_1 is known as the residue of $F(s)$ at $s = p_1$.

With the coefficients in the partial fraction expression found, we need to find the inverse Laplace transforms for terms such as $(s+a)^{-m}$. In order to do so, we first study the Laplace transform of time functions such as t^n. Firstly,

$$L\{t\} = \int_0^\infty te^{-st}\,dt = t\frac{e^{-st}}{-s}\bigg|_{t=0}^{\infty} - \int_0^\infty \frac{e^{-st}}{-s}\,dt = 0 - \frac{e^{-st}}{(-s)^2}\bigg|_{t=0}^{\infty} = \frac{1}{s^2}$$

In a similar fashion, we have

$$L\{t^2\} = \int_0^\infty t^2 e^{-st}\,dt = t^2 \times \frac{e^{-st}}{-s}\bigg|_{t=0}^{\infty} - \int_0^\infty 2t\frac{e^{-st}}{-s}\,dt = 0 + \frac{2}{s}L\{t\} = \frac{2}{s^3}$$

In general, we have

$$L\{t^n\} = \int_0^\infty t^n e^{-st}\,dt = t^n \frac{e^{-st}}{-s}\bigg|_{t=0}^{\infty} - \int_0^\infty nt^{n-1}\frac{e^{-st}}{-s}\,dt$$

$$= 0 + \frac{n}{s}L\{t^{n-1}\} = \frac{n}{s}L\{t^{n-1}\}$$

By the Theorem of Induction, we have

$$L\{t^n\} = \frac{n!}{s^{n+1}}$$

Furthermore, we have

$$L\{e^{-at}t^n\} = L\{t^n\}_{s=s+a} = \frac{n!}{s^{n+1}}\bigg|_{s=s+a} = \frac{n!}{(s+a)^{n+1}}$$

Consequently,

$$L^{-1}\left\{\frac{1}{(s+a)^{n+1}}\right\} = \frac{t^n}{n!}e^{-at}$$

Let us examine the following example.

Example 10.4 Assuming that $i(0)$ and $v(0)$ equal to zero and that the source voltage $v_s = \cos(t)$ V for the circuit in **Figure 10.2**. Find $i(t)$ for $t \geq 0$.

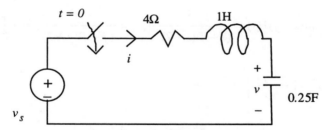

Figure 10.2 An RLC circuit

Solution: Applying Kirchoff's voltage law to the circuit for $t \geq 0$, we have

$$v_s(t) = 4i(t) + 1\frac{di}{dt} + v(t) \qquad 10.3\text{-}1$$

The voltage of the capacitor is related to the current as follows:

$$v(t) = \frac{1}{0.25}\int_{-\infty}^{t} i(t)dt$$

Taking the Laplace transform of both sides **Equation 10.3-1**, we have

$$\frac{s}{s^2+1} = 4I(s)+sI(s)+V(s)$$

$$= 4I(s)+sI(s)+4\frac{1}{s}I(s) = 4I(s)+sI(s)+\frac{4}{s}I(s)$$

Solving for $I(s)$, we have

$$I(s) = \frac{s/(s^2+1)}{4+s+4/s} = \frac{s^2}{(s^2+1)(s^2+4s+4)} = \frac{s^2}{(s-j)(s+j)(s+2)^2}$$

$$= \frac{a}{s-j} + \frac{b}{s+j} + \frac{c}{s+2} + \frac{d}{(s+2)^2}$$

Due to the fact that the coefficients in the rational function representing $I(s)$ are all real numbers, coefficients "a" and "b" for the pair of complex conjugate poles also form a complex conjugate pair. These coefficients can be found as follows:

$$a = \lim_{s\to j}(s-j)I(s) = \frac{s^2}{(s+j)(s+2)^2}\Big|_{s=j} = \frac{j^2}{(j+j)(j+2)^2} = \frac{4+j3}{50}$$

$$b = a^* = \left(\frac{4+j3}{50}\right)^* = \frac{4-j3}{50}$$

$$d = \lim_{s\to -2}(s+2)^2 I(s) = \frac{s^2}{(s^2+1)}\Big|_{s=-2} = \frac{4}{4+1} = \frac{4}{5}$$

$$c = \lim_{s\to -2}\frac{d}{ds}\left[(s+2)^2 I(s)\right] = \lim_{s\to -2}\frac{d}{ds}\left[\frac{s^2}{s^2+1}\right]$$

$$= \left[\frac{2s}{s^2+1} - \frac{2s^3}{(s^2+1)^2}\right]_{s=-2} = \frac{-4}{4+1} - \frac{-16}{(4+1)^2} = \frac{-4}{25}$$

Finally, for $t > 0$, we have the following:

$$i(t) = L^{-1}\{I(s)\}$$

$$= aL^{-1}\{\frac{1}{s-j}\} + bL^{-1}\{\frac{1}{s+j}\} + cL^{-1}\{\frac{1}{s+2}\} + dL^{-1}\{\frac{1}{(s+2)^2}\}$$

$$= ae^{jt} + a^*e^{-jt} + ce^{-2t} + dte^{-2t} = 2\operatorname{Re}\{ae^{jt}\} + ce^{-2t} + dte^{-2t}$$

$$= \frac{4}{25}\cos t - \frac{3}{25}\sin t - \frac{4}{25}e^{-2t} + \frac{4}{5}te^{-2t}$$

10.4 Theorems on Laplace Transform

With the Laplace transform properly defined, other important concepts and theorems have been developed. The concept of transfer function that is widely used in the study of linear circuits and systems is to be introduced in this section. Theorems such as the initial value and the final value theorems are also to be presented thereafter.

10.4.1 Impulse and Transfer Function

For the circuit in **Figure 10.2**, the characteristic of the circuit is governed by **Equation 10.3-1**. Taking the Laplace transform of the equation, we have

$$V_s(s) = 4I(s) + sI(s) + \frac{4}{s}I(s)$$

where $V_s(s)$ is the Laplace transform of the source voltage $v_s(t)$. Note that the equation is derived with the assumption that $i(0)$ and $v(0)$ equal to zero.

Solving for $I(s)$, we have

$$I(s) = \frac{V_s(s)}{4+s+4/s} = \frac{s}{s^2+4s+4}V_s(s) = T(s)V_s(s)$$

where

$$T(s) = \frac{I(s)}{V_s(s)} = \frac{s}{s^2+4s+4}$$

When the current $i(t)$ is considered as the output of the circuit, the function $T(s)$ is referred to as the transfer function of the circuit. The transfer function of an RLC circuit stimulated by a single source possesses the following properties:

(a) It is a proper rational function of the complex frequency s with real coefficients, i.e., the ratio of two polynomials with the degree of the numerator less than or equal to the degree of the denominator.

(b) Poles of the transfer function are either in the left-half of the complex s-plane or on the imaginary axis. They are symmetrically located with respect to the real axis.

When a transfer function $T(s)$ is strictly proper, i.e., the degree of the numerator is less than the degree of the denominator, the inverse Laplace transform of $T(s)$ represents a well-defined time function known as the impulse response.

The term impulse refers to the often-mentioned impulse function $\delta(t)$. Strictly speaking, the impulse function is not a well-defined time function. It can only be defined as an operator (specifically a functional) that maps from a function space to the real number set. The function space consists of real-valued time functions with properties such as being continuous, differentiable, ..., etc. These well-behaved time functions are referred to as test functions. Given a time function g(t), integrable but not necessary with the same properties of a test function, an operator can be defined as follows:

$$< g, f > = \int_{-\infty}^{\infty} g(t)f(t)dt$$

where the improper integration exists for any test function $f(t)$ due to the good nature of a test function.

When a test function $f(t)$ is operated on by the impulse, we get the real number $f(0)$. Such an operation is often denoted as follow:

$$< \delta, f > = \int_{-\infty}^{\infty} \delta(t)f(t)dt = f(0)$$

Although the misuse of the notation of integration is apparent in the equation above, it does give rise a sense of familiarity. Furthermore, other well-known properties concerning the impulse can easily be associated with the improper integration and be shown in the following.

$$< \delta_\tau, f > = \int_{-\infty}^{\infty} \delta(t-\tau)f(t)dt = f(\tau)$$

$$< \delta, \quad 1 > = \int_{-\infty}^{\infty} \delta(t)dt = 1$$

Furthermore, the Laplace transform of the impulse can be defined as follows:

$$< L\{\delta\}, f > \stackrel{def}{=} \; <\delta, L\{f\}>$$

$$= <\delta, \int_{-\infty}^{\infty} f(t)e^{-st}\,dt> = \int_{-\infty}^{\infty} f(t)e^{-st}\,dt\bigg|_{s=0} = \int_{-\infty}^{\infty} f(t)dt = <1, f>$$

In the last equation, the test function $f(t)$ is chosen among those functions which are zero-valued for $t \leq 0$. From the equation, it is claimed that the Laplace transform of the impulse is the constant function $G(s)=1$. Hence, we have

$$I(s) = T(s)V_s(s) = T(s)$$

if the input (the voltage source) is an impulse. It is why the inverse Laplace transform of $T(s)$ is referred to as the impulse response.

10.4.2 Convolution Integral

The concept of transfer function has been widely used in the study of circuits and systems. Its popularity can be attributed directly to the operation known as the convolution on two time functions. For two piece-wise continuous functions $g(t)$ and $f(t)$ that maps from $[0,\infty)$ to the real number set R, the convolution gives rise to a new time function defined as follows:

$$f * g(t) = \int_0^t f(t-\tau)g(\tau)d\tau$$

With a simple change of variable, we have

$$f * g(t) = \int_0^t f(t-\tau)g(\tau)d\tau = -\int_t^0 f(x)g(t-x)dx$$

$$= \int_0^t f(x)g(t-x)dx = g * f(t)$$

When the Laplace transform of $f * g(t)$ is performed, we have

$$L\{f * g(t)\} = \int_0^\infty \left[\int_0^t f(t-\tau)g(\tau)d\tau \right] \cdot e^{-st}\,dt$$

$$= \int_0^\infty \left[\int_0^\infty f(t-\tau)g(\tau)u_0(t-\tau)d\tau \right] \cdot e^{-st} dt$$

$$= \int_0^\infty \int_0^\infty f(t-\tau)g(\tau)u_0(t-\tau)e^{-st} dt d\tau$$

$$= \int_0^\infty \left[\int_\tau^\infty f(t-\tau)e^{-st} dt \right] \cdot g(\tau)d\tau = \int_0^\infty \left[\int_0^\infty f(x)e^{-s(x+\tau)} dx \right] \cdot g(\tau)d\tau$$

$$= \int_0^\infty F(s)e^{-s\tau} g(\tau)d\tau = F(s)G(s)$$

Example 10.4 (continued) Find the impulse response for the circuit shown in **Figure 10.2.**

Solution: Previously, we have found the transfer function of the circuit. Taking the partial fraction expansion of the transfer function, we have

$$T(s) = \frac{I(s)}{V_s(s)} = \frac{s}{s^2 + 4s + 4} = \frac{1}{s+2} + \frac{-2}{(s+2)^2}$$

Let $h(t)$ denote the inverse Laplace transform of $T(s)$. We have

$$h(t) = L^{-1}\{T(s)\} = e^{-t} - 2te^{-t}$$

for $t > 0$.

With the impulse response, the current due to a particular source voltage can be found via the convolution integration as follows:

$$i(t) = h * v_s(t) = \int_0^t h(t-\tau)v_s(\tau)d\tau$$

for $t > 0$.

Although convolution offers a nice formulation of the output of a circuit stimulated by any given input, it is often tedious to find the result through integration. With the Laplace transform, the complex integration of convolution is turned into a mere multiplication. When designing a linear circuit or a system, multiplication of Laplace transforms offers a much simpler alternative to integration. After the design is done, taking the partial fraction expansion of a

Laplace transform is the preferred method for finding the time function if it is desired.

10.4.3 Initial-Value and Final-Value Theorems

When designing a circuit using the Laplace transform, the transform of the circuit response can easily be calculated. To spot check the time-domain response, it may be desirable to calculate directly a particular value of the time function without carrying out the elaborate partial fraction expansion of the transform. There are two particular values of a time function $f(t)$ that can be found in such a manner: the initial value $f(0)$ and the final value $f(\infty)$. If $f'(t)$ is piecewise continuous, we have

$$L\{f'(t)\} = \int_0^\infty f'(t)e^{-st}\,dt = -f(0) + sF(s) \qquad 10.4\text{-}1$$

Rearranging **Equation 10.4-1**, we have

$$f(0) = \int_0^\infty f'(t)e^{-st}\,dt - sF(s)$$

Since the right hand side of the equation must be a constant for all the complex frequency s in the region of convergence, let us find out what happen when s approaches to ∞ along the positive real axis. We have

$$\lim_{s\to\infty} \int_0^\infty f'(t)e^{-st}\,dt = \int_0^\infty f'(t)\lim_{s\to\infty}e^{-st}\,dt = 0$$

where the interchange of the two limit processes is assumed to be permissible and can be proven. Consequently, we have

$$f(0) = \lim_{t\to\infty} sF(s) \qquad 10.4\text{-}2$$

Equation 10.4-2 is known as the initial-value theorem.

On the other hand, if the origin is contained in the region of convergence, we can make s approach zero along the positive real axis. We have

$$\int_0^\infty f'(t)e^{-st}\,dt\bigg|_{s=0} = \int_0^\infty f'(t)\,dt = f(t)\bigg|_{t=0}^{\infty} = f(\infty) - f(0)$$

Consequently,

$$f(\infty) = \lim_{s \to 0} sF(s) \qquad\qquad 10.4\text{-}3$$

Equation 10.4-3 is known as the final value theorem. Note that the origin is contained in the region of convergence if all poles of F(s) are located inside the left half of the complex s-plane. This condition can actually be lessened slightly by checking whether the poles of $sF(s)$ are inside the left half plane before the theorem is applied.

Example 10.5 The Laplace transform of a function $f(t)$ is found to be

$$F(s) = \frac{s^2 + 2}{s(s^2 + 4s + 4)}$$

Find the final value of the function, $f(\infty)$, if it exists.

Solution: Since both poles of $sF(s)$ are at s = −2, the final value theorem is applicable and we have

$$f(\infty) = \lim_{s \to 0} sF(s) = \lim_{s \to 0} \frac{s^2 + 2}{s^2 + 4s + 4} = \frac{2}{4} = \frac{1}{2}$$

Example 10.6 The Laplace transform of a function $g(t)$ is found to be

$$G(s) = \frac{s}{(s-1)(s+1)}$$

Find the final value of the function, $g(\infty)$, if it exists.

Solution: Since one pole of $sG(s)$ is at s = 1 which is in the right half plane, the final value theorem is not applicable. In other words, $g(\infty)$ doesn't exists. However, the following limit does exist

$$\lim_{s \to 0} sG(s) = \lim_{s \to 0} \frac{s^2}{(s-1)(s+1)} = 0$$

10.5 Solving the State Equations with Laplace Transform

The state equations of an RLC circuit are a collection of some linear simultaneous first-order differential equation with constant coefficients. Through Laplace transform, these equations are turned into simultaneous algebraic equations. The Laplace transforms of the state-variables can thus be found through algebraic operation. State-variables themselves can then be found through partial fraction expansion. With this kind of approach, to find the differential equation that governs a particular circuit quantity plus appropriate initial conditions for that circuit quantity is no longer necessary. Let us examine **Example 10.6**.

Example 10.6 The circuit shown in **Figure 10.3** is the same one as the one in **Figure 5.10**. Find the voltage across the capacitor and the current through the inductor.

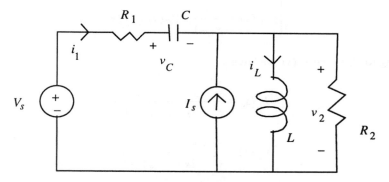

Figure 10.3. An RLC circuit

Solution: In **Chapter 5**, the state equations of the circuit are found to be

$$\frac{di_L}{dt} = \frac{1}{L(R_1 + R_2)}(R_2V_s - R_2v_C + R_1R_2I_s - R_1R_2i_L)$$

$$\frac{dv_C}{dt} = \frac{1}{C(R_1 + R_2)}(V_s - v_C - R_2I_s + R_2i_L)$$

Let us further assume the following.

$$R_1 = R_2 = 1\Omega, \quad L = 0.5\text{H}, \quad C = 0.5\text{F}$$
$$V_s(t) = \cos(t), \quad I_s(t) = 0$$
$$v_C(0) = 0, \quad i_L(0) = 0$$

With these assumptions, the state equations become

$$
\begin{cases}
\dfrac{di_L}{dt} = V_s - v_C - i_L \\[4mm]
\dfrac{dv_C}{dt} = V_s - v_C + i_L
\end{cases}
$$

Taking the Laplace transform of the state equations, we get

$$
\begin{cases}
sI_L(s) = \dfrac{s}{s^2 + 1} - V_C(s) - I_L(s) \\[4mm]
sV_C(s) = \dfrac{s}{s^2 + 1} - V_C(s) + I_L(s)
\end{cases}
$$

Solving for $I_L(s)$ and $V_C(s)$, we have

$$
I_L(s) = \frac{s}{s^2 + 2s + 2} \cdot \frac{s}{s^2 + 1}
$$

$$
V_C(s) = \frac{s + 2}{s^2 + 2s + 2} \cdot \frac{s}{s^2 + 1}
$$

The time functions $i_L(t)$ and $v_C(t)$ can be found through partial fraction expansion.

Problems

10.3-1 Determine the values of the constants in the partial fraction expansions of the following functions:

(a) $\dfrac{(s+2)(s+3)}{(s+1)(s^2 + 2s + 2)} = \dfrac{a}{s+1} + \dfrac{b}{s+1+j} + \dfrac{c}{s+1-j}$

(b) $\dfrac{s^2 + s + 1}{(s+1)^2 (s+2)^2} = \dfrac{a}{s+1} + \dfrac{b}{(s+1)^2} + \dfrac{c}{s+2} + \dfrac{d}{(s+2)^2}$

10.3-2 Find the partial fraction expansion of the following functions:

(a) $\quad V(s) = \dfrac{s^2 + 1}{s(s+1)(s+2)[(s+1)^2 + 1]}$

(b) $\quad V(s) = \dfrac{s^4 + 2s^2 + 1}{s(s+1)^3(s+2)}$

(c) $\quad V(s) = \dfrac{s^5 + s^2 + 1}{s(s+1)^2(s+2)[(s+1)^2 + 1]}$

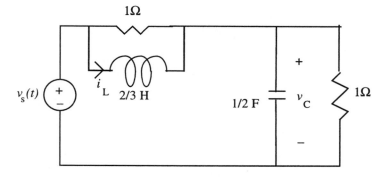

Figure 10.4. An RLC circuit

10.4-1 Apply the initial-value and the final-value theorems, if appropriate, to the following functions:

(a) $\quad F_1 = \dfrac{s^2 + 1}{s(s+1)(s+2)}$

(b) $\quad F_2 = \dfrac{s^2 + 6s}{(s+1)(s^2 + 4)}$

(c) $\quad F_3 = \dfrac{(s^2 + 1)(s+2)}{s^3(s+1)}$

10.4-2 In the circuit shown in **Figure 10.4**, $v_s(t)$ is a unit step time function. The initial voltage across the capacitor is $1V$ and the initial current through the inductor is $0\,A$. Find the voltage across the capacitor using the Laplace Transform.